T3-BOC-470

The Fifth Generation Computer

The Fifth Generation Computer

The Japanese Challenge

Tohru Moto-oka

Department of Electrical Engineering,
University of Tokyo, Japan

and

Masaru Kitsuregawa

Department of Electronic Computation,
The Institute of Industrial Science,
University of Tokyo, Japan

Translated by
F. D. R. Apps, *Ingatestone Translations*

JOHN WILEY & SONS
Chichester · New York · Brisbane · Toronto · Singapore

Library of Congress Cataloging in Publication Data:

Motooka, Tōru, 1929–
The fifth generation computer.
Translation of: Daigo sedai konpyuta.
Includes index.
1. Electronic digital computers.
I. Kitsuregawa, Masaru. II. Title.
QA76.5.M64913 1985 0004 85-12290

ISBN 0 471 90739 1 (pbk.)

British Library Cataloguing in Publication Data:

Moto-oka, T.
The fifth generation computer.
1. Electronic digital computers.
I. Title II. Kitsuregawa, Masaru
001.64 QA76.5

ISBN 0 471 90739 1

Typeset by Acorn Bookwork, Salisbury, Wiltshire
Printed and bound in Great Britain at The Bath Press, Avon

Contents

Introduction

The Fifth Generation Computer Project has attracted widespread interest, both at home and abroad, and the project has got off to a good start as hoped. The reaction from Europe and the USA has been greater than expected, and augurs well for its future.

The end of the 1970s was a period of worldwide economic slump, with a scaling-down of expectations regarding the benefits to be had from technology. This in turn was reflected in a diminution in the number of applicants for university courses in science and engineering. Against this background, the Fifth Generation Computer Project has offered fresh hope to society. Together with robotics and micro-electronics, it represents one of the first steps towards the age of high technology.

The aims of the Fifth Generation Computer Project are to open up the new fields of application that are referred to by the terms 'knowledge processing systems' and 'applied artificial intelligence systems', and to create a user-friendly interface between computer and user. Aiming as it does to combine the power of a wide range of technologies, including VLSI, computer architecture, software, and artificial intelligence, the Fifth Generation Computer Project has many different aspects, and so may give the impression of being difficult to understand.

Unlike many previous projects in Japan, this Fifth Generation Computer Project has not been set up with any very well-defined objective; only the direction of development has been clearly given and little has been explicitly laid down regarding long-term attainment. The intention is to move the project forward steadily by setting a succession of short-term goals. Some people therefore consider it as a goal-searching project.

The general outline of the Fifth Generation Computer Project has already been made widely known by the book by Professor Feigenbaum *et al.*, but, in spite of the fact that this is a project which has been initiated in Japan, there has so far been no book written by a Japanese giving an explanation of the project for the general reader. Although I doubted my suitability for the task, I therefore accepted, at the strong urging of Messrs S. Kobayashi and H. Katayama of the Iwanami Shoten Publishing Company, the challenge of writing an introduction to the Fifth Generation Computer Project.

To make the text rather easier to read, the method which we adopted was that I dictated the first version of the text, which was then revised by Dr Kitsuregawa, and I then again checked and revised his text. Though the text may contain a certain amount of dogmatism and personal prejudice, we shall be pleased if it is found to be of some service in giving the general reader some understanding of future computers.

Tohru Moto-oka
Summer 1984

Chapter 1

The revolutionary fifth generation computer

THE ANNOUNCEMENT TO THE WORLD OF THE concept of the fifth generation computer took place at the 'International Conference on the Fifth Generation Computer' held in Tokyo from the 19th to the 22nd of October 1981. The fifth generation computer project was initially planned some three years earlier.

In this chapter we look at how this project started, and the response to it from various parts of the world.

Origin of the project

It was April or May of 1979 when I (Moto-oka) was approached by JIPDEC (Japan Information Processing Development Association), an extra-governmental organization of MITI (Ministry of International Trade and Industry). It seems that there had been some discussion in the electrotechnical laboratory (ETL) and other fundamental research institutions of the Ministry, and also in the Ministry itself for the previous year or so, but it was from 1979 that the Ministry requested a survey relating to the fifth generation computer from JIPDEC.

At this point, MITI's viewpoint was that although Japan's computer industry had been developing up to that point, it was not clear what Japan, as a nàtion, should be doing in this field in the 1980s. Already computer hardware of a quality needed for export was being produced, but nowhere in the world was there a clear development target for the 1990s. The call from MITI was for about two years' study to consider the likely state of society, both in Japan and in other countries, in the 1990s, to investigate the role of the computer in that decade, to point out any problems requiring national action, and to suggest an approach to research problems.

Initiation of the survey committee

In response to this requirement, a 'Fifth generation computer survey committee', of which I (Moto-oka) was privileged to be chairman, was set up within JIPDEC. Since I was just fifty, 1990 will mark my retirement from university and I accepted gladly.

Since computers of the 1990s can be approached from three directions, we set up three special-interest groups, one for each field. The first was to consider what kind of computer the society of the 1990s would require, and the responsibility for this was taken by Hajime Karatsu, who at the time belonged to the Matsushita Tsushin Company. The second group was to look at architecture from the computer technology viewpoint, and this was led by Hideo Aiso, Professor of Electrical Engineering at Keio University, while the third group, considering basic theory, was led by Kazuhiro Fuchi, at the time director of the pattern information section of the electrotechnical laboratory. Fig. 1 shows the organization of the 1980 survey committee. There were minor differences in the 1979 structure, in that there was no task group, and the systems group was replaced by a social environment subcommittee.

The survey committee began discussion from the understanding that the fifth generation computer would be substantially different in character from any existing computers. The basic theory subcommittee considered quite fundamental research questions with the help of mathematicians, linguists, and so forth. The social environment subcommittee was participated in by a wide range of computer users, from banks, trading companies, and a variety of manufacturing industries, providing many different starting points for discussion.

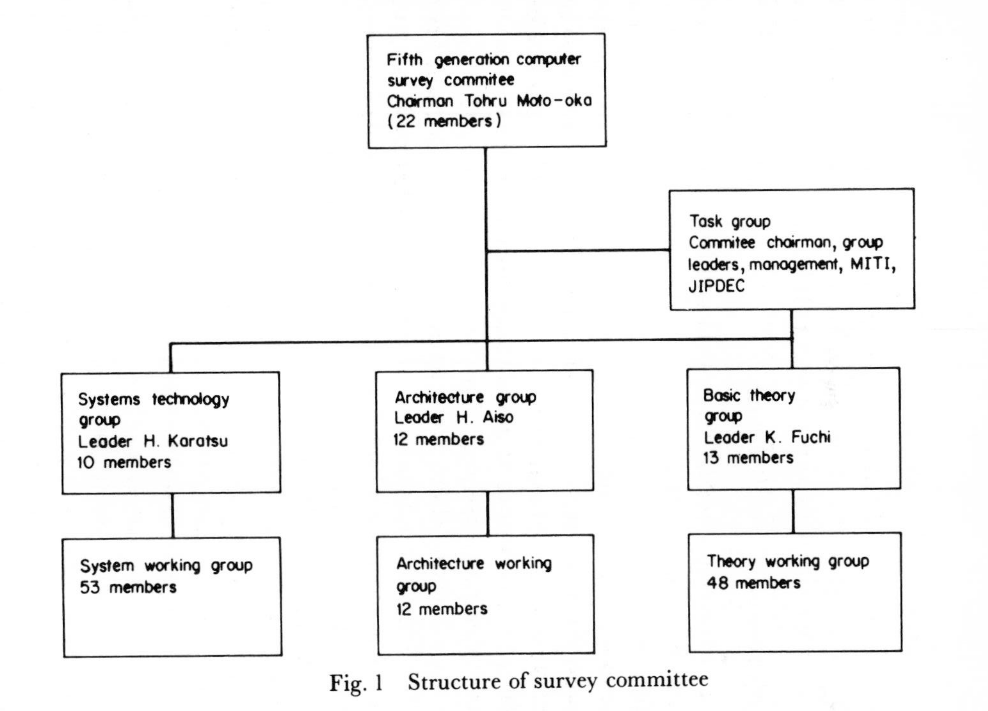

Fig. 1 Structure of survey committee

In a two-year period, over a hundred people participated in one way or another in the committee's efforts; there were more than a hundred meetings of various forms, and we were able to hear a wide range of opinions. Thus we arrived at a target for the fifth generation computer.

International conference

As mentioned above, this project was first announced at the International Fifth Generation Computer Conference held in Tokyo in October 1981, but before that date there had been workshops and so forth to hear the opinions of international scholars; through JIPDEC there were discussions with research institutions of the principal countries of Europe and America, and some information was obtained from a survey visit to America.

In addition, there were invitations to participate from MITI to American and European governments, and because the project had already become well known abroad, more than eighty overseas participants attended the Fifth Generation Computer Conference.

Before this conference I spent a month with Kazuhiro Fuchi, touring America, Germany, France, and Great Britain, in preliminary discussions with participants. We had many discussions—in particular with Colmeraur in France, who has played such an important part in the practical implementation of the programming language PROLOG, and in America with Feigenbaum, the proponent of expert systems. The general reaction at that stage was that the project was very ambitious, and some criticisms were heard to the effect that our goals were not sufficiently concrete.

In addition, there seem to have been times when Michael L. Dertouzos, the head of the artificial intelligence department at MIT (Massachusetts Institute of Technology), had

Fig. 2 First International Conference on the Fifth Generation Computer. Many attendees from overseas

doubts as to whether their own current plan for an artificial intelligence project, which was quite similar to the fifth generation computer project, was in fact being stolen. At the international conference, however, we were able to show that the Japanese plan is actually wider in scope and larger in scale than the MIT project. Later I was able to meet and talk with Dertouzos, was invited to MIT, and was I believe able to resolve this misunderstanding.

Internationalization of the project

Until the international conference, this project had been discussed principally from a domestic Japanese point of view, but on the other hand, one purpose of the conference was to urge that the project should not be restricted to Japan but should be a matter for international cooperation, and this is the point of view from which overseas workers were looking at the project. Perhaps we wanted a wider meaning outside the normal bounds of international conferences.

For example, practically no discussion had taken place up to that time of the meaning of the fifth generation computer project to the third world, but in order to earn the label 'world project', it was felt that such problems had to be discussed. Again, we were made to realize how much different perspectives can change perception of a situation: for example, rather than asking for international cooperation, Japan could be seen as seeking to extract expertise from experts collected from around the world.

European countries such as Britain, Germany, and France each sent a number of representatives, principally from government positions. For this reason, compared with usual international conferences, the results of the conference permeated relatively quickly to these countries.

The British response

In Britain, as a result of this international conference, a committee was set up the same year (1981) with John Alvey from British Telecom as chairman, to determine what should be Britain's response to the Japanese fifth generation project. It produced a report ('The Alvey Report') in the autumn of the following year (1982) which urged that it was necessary for Britain too to provide government support for computer research. The project started with the provision from about June 1983 of a budget of 500 million dollars over five years.

The range of the project was somewhat wider than that of the Japanese fifth generation project, including research on topics from VLSI development to knowledge processing along the lines of the Japanese project, and also on topics such as intelligent man–machine interfaces. It involves accordingly both industry and universities.

The Japanese project is, of course, a national project, but on the other hand comes in practice under the aegis of MITI, whereas the British project is truly a national project, involving as it does first the Department of Trade and Industry, and also the Department of Education and the Ministry of Defence. From the Japanese point of view it is an enviable arrangement, including an education structure and also the provision of a computer network for research purposes.

American research

As the American response to the Japanese plan, a group of manufacturers centred around Control Data Corporation, though not including IBM, set up a research consortium called MCC (Microelectronics and Computer Technology Corporation) in which a large number of computer manufacturers and semiconductor manufacturers are participating.

Moreover, since 1983 the Defense Applied Research Projects Agency (DARPA) of the US Department of Defense has started a programme for strategic programming research, because of a realization of the need for national action on this aspect of research. This project is being promoted on the basis of starting research for a computer for artificial intelligence from 1990, and involves both university and industry participation. DARPA is contributing 100 billion dollars to the research costs.

West German, French, and EC plans

West Germany was rather later entering the field, but in 1984 decided on a national budget of DM 3,600 million over five years on information-related leading edge technology development. Of this budget DM 900 million is related to data processing, including knowledge base systems and new computer architectures. This is the first project to impinge on the Japanese fifth generation computer project; but it concentrates on the most pressing problems and postpones the response to the fifth generation computer until the next phase in a few years' time. At present, preparations for the project seem to be well under way.

In France also, a committee called SICO (Systèmes Informatiques de la Connaissance) has been set up at the national laboratory for information processing research, INRIA, and this committee is debating the steps necessary in France to promote research related to artificial intelligence. Its report was published in June 1983, and led to a request to the government for the necessary budget.

The European Community has produced a plan called ESPRIT (European Strategic Plan for Research in Information Technology), which is just coming into effect. This is centred on twelve large computer companies in Britain, Germany, and France, and also in Italy and The Netherlands, and aims to prevent Europe from falling behind Japan and America in the information processing field by encouraging European cooperation in exploiting its research potential. For this project, 1983 was the preparatory phase with only a pilot scheme in operation; but from 1984 the project will start officially, and over five years will receive 1,600 million ECUs (European Currency Units, slightly less than one American dollar each). Each project involves both industry and universities or other research institutes from at least two different countries, and half of the cost is contributed by the industrial participants. Already ICL from Britain, Bull from France, and Siemens from West Germany have proceeded with the setting up of a joint research institute in knowledge engineering in southern Germany.

Fig. 3 Participants at the international conference

Worldwide cooperation

As described above, fifth generation computer research has reached the stage of application to the common goals of the major countries of the world. The aim of the fifth generation project has always been, not merely to achieve independent progress for Japanese research, but for the peoples of the world to approach the same objectives from their separate standpoints. In other words, we have arrived at a very convenient world situation.

A drawback, though, of the financial involvement in the American case of the Department of Defense is the danger of information not being shared with other countries, which would impair the aim of global cooperation. Since this is basic research, it would be preferable for results to be shared between countries, in an effort to develop together a computer for the benefit of future generations, rather than to be kept as national secrets.

One reason for holding the international conference was the hope that, in a project as difficult as the fifth generation computer, for which Japan alone has insufficient resources, help would be available from other countries. This was mooted at an early stage, but international cooperation to develop a new computer is not completely straightforward.

For example, in an international project to build a new aircraft, Japan might build the fuselage, and Britain the engines; but a computer cannot be so easily cut into sections. At present, in the design sequence, we are at a more fundamental stage of research.

The Japanese view is that for the time being the various countries should continue with their research programmes in parallel, but that researchers should be exchanging information on an individual basis. Then when a country develops a strength in a particular field, there should be an international division of labour.

Journals for information exchange

It is very important, for research to proceed successfully, to have an international exchange of information. Since most publications in Japan are in Japanese, they are not well known in other countries, and the result of this is either that Japanese research is not properly evaluated, or that the impression is given that Japan is seeking to hide research results.

Because it was realized that this could be a problem with the fifth generation computer project, it was decided to publish an English language journal, based in Japan; it is a quarterly called *New Generation Computing* (Editorial board Chairman, Tohru Moto-oka; Vice-chairman Kazuhiro

Fuchi), and has been published since 1983. Although the editorial committee is based in Japan, it will be helped by a team of overseas editors, including R. Kowalski from Britain, T. Winograd from the USA, and K. Berkling from Germany, and the publishers, Springer, who have a high reputation for academic publishing, will ensure worldwide availability.

As part of the same effort, in Europe Elsevier of the Netherlands are producing a newsletter called *Fifth Generation Computing*. This will follow developments in various countries, and will try to show research trends related to the fifth generation computer, and it is therefore hoped that it may later assume the format of a journal.

Towards a second international conference

The fifth generation computer project in Japan began officially in 1982 and is planned to extend for ten years. The first period of three years ends in March 1984. In order to publish the results of that initial period, and to give the international computing community a chance to air its views, a second international conference is to be held in Tokyo in November 1984.

This time I (Moto-oka) shall again be chairing the conference, but it is hoped that researchers from all over the world will be attending; already some 160 papers have been submitted from 24 different countries. Time constraints will mean that not all of these can be presented, but as well as presenting results from ICOT (Institute for New Generation Computer Technology) in Japan, the conference will be a wider forum for related research being carried out all over the world. It is hoped that this conference will be one opportunity to advance fifth generation computer research.

Chapter 2

The concept of the fifth generation computer

IN THIS CHAPTER, WE SHALL EXPLAIN WITH WHAT AIMS, and against what background, the concept of the Fifth Generation Computer was arrived at.

The problems that face present-day computers

The Fifth Generation Computer Project was conceived with a consciousness of the problems that face present-day computers. Probably the thinking in MITI was: now that we have nurtured a Japanese computer industry that is capable of meeting the challenge of IBM, what should we do next?

Computer researchers had the same thought. In my own case, the year in which I graduated from university (1952) witnessed the beginning of serious study of electronic computers involving the Department of Engineering at The University of Tokyo (the computer was known as TAC, standing for 'Tokyo Automatic Digital Computer'). Over the ensuing thirty years, computers advanced from the first to the third generation, and the computer industry or computer technology has now assumed vast proportions. And an enormous amount of software to enable these computers to be used has been accumulated.

Because of this, there has been a tendency to think that further changes in computer architecture are not possible. However, it has gradually become clear that there are limits to the extent to which the performance of computers can be raised without changing their architecture. (In its narrow sense, the term 'architecture' means the set of computer instructions and their meanings, i.e. the properties and characteristics of the computer as seen by the programmer. But in a broader sense it is often used to mean the entire logic structure of the hardware system of the computer. Here the term is to be understood in its broad sense.)

On the other hand, new techniques such as VLSI are now

becoming available. This new technology is not very well utilized in present computers, and indeed the effective utilization of this technology does present problems with existing architecture. There is therefore a strong feeling that a fundamental change in computer technology is required. We shall discuss this in more detail in the next chapter.

The software crisis

Another problem is the 'software crisis'. This means the very great difficulty of constructing large software—or, once the software is constructed, the very considerable amount of labour that is required to improve and maintain it so that it always meets requirements.

For example, the operating system, which is the software that makes the computer operate as a system, consists of a huge program of several hundred thousand steps, and it is not uncommon for even a moderate applications program to require anything from several thousand to several tens of thousands of steps.

To solve this problem, much study has been given to the questions how software productivity can be raised and how software maintenance can be facilitated. This is called software engineering. However, to make use of such studies in the practical field, we need to have another look at the computer itself.

Present computers are not good at non-numerical processing

Computers were originally constructed to perform numerical calculations. However, since they were first constructed, many experts have pointed out that they can be used for purposes other than numerical computation. The operation of computers and of the brain were thought of as being similar and indeed in the early designs terms were often used on the analogy of the nervous system.

Examples of ways in which computers can be utilized, apart from numerical calculation, are: making a computer carry out a translation; reading text; or construction of a robot. When we attempt to use computers to handle objects other than numerical values, or images or drawings, or attempt to deal with language, we come up against the problem that the capability of the computer is very low.

For example, about twenty years ago, an immense amount of research was carried out worldwide concerning the problem of 'machine translation', i.e. getting a computer to perform a translation. As a result people found out that translation is not such a simple task after all. What brought this research to a sudden halt was the realization that little progress could be made with simple grammatical elucidation and word-by-word translation. Genuine automatic translation will require not merely mechanical symbol processing but understanding of meaning, which involves giving the computer commonsense like a human being.

Basic research was nevertheless steadily continued into such artificial intelligence problems as understanding of natural language and robotics. This has led to a fairly good understanding as to how the language problem should be handled by computers.

Nevertheless problems such as understanding of natural language still cannot be handled very efficiently with present-day computers. This is because present-day computers are numerical computation computers. Though they have a high degree of calculating power in terms of arithmetic and floating decimal point calculations etc., for non-numerical data they only have the very basic functions of taking the logical sum (AND) or logical product (OR).

Processing non-numerical data by means of combinations of such very basic functions is extremely inefficient. This is a big hindrance to artificial intelligence research aimed at understanding of natural languages, for example.

If a computer suited to processing such non-numerical data could be developed, new fields of application such as understanding of natural language might be rapidly opened up. This is not to say that a computer could be given a human-like understanding of language at one fell swoop. The first-stage objective is to give a computer the ability to understand comparatively simple language. Even this could provide support for human beings in many fields of application. Once a start has been made in a particular field of application, there is no doubt that there will be steady progress in that direction, once the consciousness that that type of problem can be handled by a computer has been created.

Image creation for the fifth generation computer

The various subcommittees of the Fifth Generation Computer Survey Committee have carried out in-depth studies regarding the image and role of the fifth generation computer.

The 'Basic Theory' subcommittee has been concerned with requirements and the possibilities that may be expected to be opened up by the new computer, as seen from the standpoint of artificial intelligence or software engineering.

The 'Computer Architecture' subcommittee, on the other hand, has been concerned with establishing what sort of computer should be built and what things it should be made to do if the new VLSI technology is to be fully utilized. Essentially there were two problems. Firstly, with VLSI techniques, costs can be brought down very low if large numbers of identical units can be used. The first problem is therefore how to construct a computer using a large number of identical units. The second problem is, since a very rapid drop in hardware costs can be achieved by VLSI techniques, what sort of computer architecture should be used to incorporate hardware having higher-level functions?

As already mentioned, existing computers are bad at processing non-numerical data. One approach to a solution of the first problem is distributed processing or parallel processing carried out by a large number of processors, and an approach to the second problem is to perform such non-numerical processing as association, inference, and learning by hardware functions.

For computers to be used in a large number of fields, computers suited to these various fields have to be designed and made available. Recently this has become economically feasible.

The 'social environment' subcommittee has been concerned with the question of what fields of application must be considered if the range of application of computers is to be increased in the future. Their conceptual approach has been to summarize the demands of society by posing the question of how computer technology can contribute to overcoming the bottlenecks to achieving the 'ideal society', envisioned as a target for the 1990s. This has led them to focus on several questions.

The role of the fifth generation computer in society

The first question is how computers can be used to raise productivity in current low-productivity fields. In the case of Japan, considerable progress has been achieved by the use of numerical control of machine tools, production robots, and computer control of factory processes and production lines (Fig. 4).

That is, the productivity of secondary industry has been greatly increased by using computers. However, computers are scarcely used at all in primary industries such as agriculture or fishing, or tertiary industries typified by offices etc., so productivity has not been raised. Computers will have to be produced that can be used in these fields.

The second question is, given that world resources are finite, how can computers be used to assist in problems of shortage of energy and resources? This has only recently come to be a problem.

This also is mainly a problem of computer control. Many

Fig. 4 Robot production line at the FANUC Company

different applications for computers may be considered, such as simulation of atomic reactors, simulation of various phenomena in society, and control of power stations. Supercomputers performing numerical calculations are very important for such purposes. Construction of expert systems using computers is also important for such problems as prospecting for petroleum. One approach to such problems is use of computers designed for artificial intelligence. Economizing on resources will be a pressing problem for humanity in the twenty-first century, and computer technology will be indispensable for this purpose.

The third question is how computers can contribute to the international environmental problem. For Japan, which is poor in natural resources, the information industry, revolving around computers, is essential for guaranteeing international competitiveness. It is also a typical field in which Japan can contribute to the world as a whole. For Japan to live in the world it has to contribute internationally in various forms. One way of doing this may be to contribute through computer technology and computer applications technology.

The chief bottleneck that impedes Japan's international contribution is the language problem. Because of the language barrier the Japanese cannot be fully active internationally. I think that the Japanese people may be able to make a greater contribution internationally if this language barrier can be even partially overcome by devising a computer that can perform translation and interpretation, as mentioned earlier.

Fourthly, there is the world problem of an ageing population, that is, an increase in the proportion of old people in society. Japan in particular is moving to an ageing society more rapidly than regions such as Europe (see Fig. 5). It is

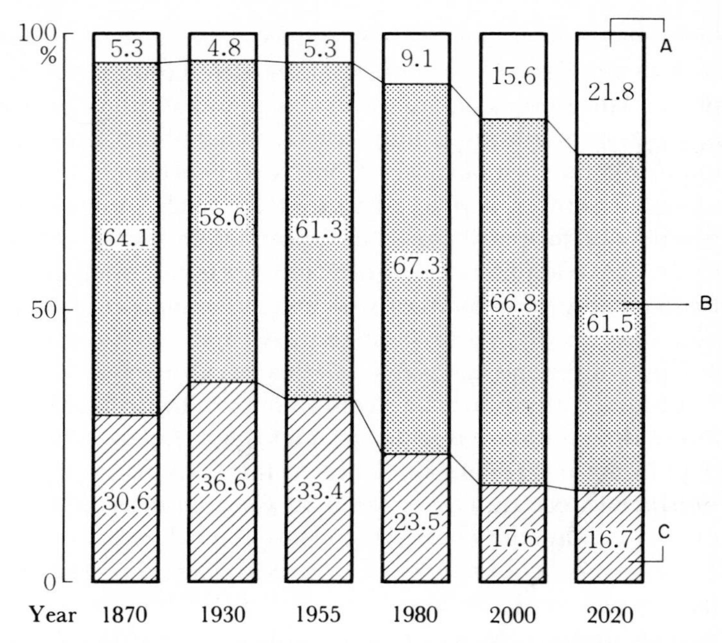

Fig. 5 Change in age structure (Japan). A: aged population (65 and over); B: population of productive age (15–64); C: child population (14 and below)

forecast that by the 1990s more than 12 per cent of the population of Japan will consist of people upwards of 65 years old. Japan will have to make in about 10 years the change to an ageing society that took about 100 years in Europe and the USA. This social transition will involve many factors which it is difficult to imagine being satisfactorily managed without the use of computers. Computers may be expected to make a considerable contribution in medical treatment systems of various kinds, mobility aid systems for disabled people, and distributed processing systems enabling part-timers to work at home without going to the office. Such questions are being examined by the social environment sub-committee.

The computer in the 1990s

The appearance of the fifth generation computer in the 1990s will not of course mean the disappearance of the ordinary computers that are being used now.

What will be the picture regarding computer use in the 1990s? Existing computers will still be being used in the field of office data processing or office computation. However, whereas at present the main application of computers in office computation is in data processing, in the 1990s, information management will become much more important. Information systems will be constructed at a world level, making available the latest most accurate information by means of a computer network constructed by linking, through communication circuits, databases in many different places.

Further use will be made of microminiaturized computers as constituent elements of systems of many different types, as in domestic electrical equipment and in the use of microcomputers in automobiles. This has already begun to happen.

Also, as mentioned previously, there will of course continue to be a strong demand for computers for numerical computation, and supercomputers performing numerical computation will be used to simulate systems of various kinds. The time is approaching when computer simulation will be used to perform much of the work that is now carried out by pilot plants and various tests using models. Apart from this, an appreciable proportion of the experiments that are carried out in physics and biochemistry will be replaced by computer simulation, and simulation of social phenomena will become common. In the field of engineering, more precise designs will be achieved by using computers.

Apart from such conventional types of computers, new computers for utilizing artificial intelligence techniques will be used in the 1990s. The main objective of such new computers will be to raise the intellectual capabilities of human beings by assisting intellectual activity in many different ways. Thus they will assist our struggle to achieve a better society and explore new worlds. We call systems having such an objective 'knowledge information processing systems'. In its narrowest sense, the 'fifth generation computer' means a computer intended for such knowledge information processing.

Knowledge information processing systems are expected to assist people by performing some of the activities such as inference and association required in the processes of decision-making and diagnosis of various systems and by acting as experts to perform a wide range of human intellectual activities, specifically work such as judgement and design. It is anticipated that such systems will have to have the ability to understand language and images, and will progressively increase in learning ability. To assist human intellectual activity, they will have to have a man–machine interface that is natural for people. (This means the ability to exchange information between people and such systems using voice or images.)

As mentioned earlier, however, the computers that will be used in the 1990s will also include information systems combining databases and computer networks, and supercomputers for scientific and technological computation, as well as large numbers of microcomputers used as system elements. It is anticipated that computers of all these different kinds will be in use in the 1990s.

Of these various types of computer, it is knowledge information processing systems that are being promoted as the Fifth Generation Computer Project.

The reasons for this arise from a consideration of the following. NTT (Nippon Telegraph and Telephone) and others have already made considerable study of computer networks, and are starting to put this into practice. And so far as supercomputers for numerical calculations are concerned, a large-scale project separate from the Fifth Generation Computer Project was begun in Japan in 1982. Development of microcomputer applications is being vigorously carried out by the various enterprises concerned.

Considering these facts, it was concluded that knowledge information systems represented the most suitable topic that should be tackled in the 1980s to enable this country to meet the challenge of the 1990s.

Chapter 3

Problems with present computers

FROM THE CREATION OF THE FIRST ELECTRONIC computer ENIAC in 1946 to present-day computers, hardware has evolved from thermionic valves through transistors and integrated circuits to LSI technology, and there have been various advances in software. Many problems remain, however, with current computers. In this chapter we consider some of these problems, and take a historical look at the development of computers, and the directions in which computers are now moving, in particular the fifth generation computer.

The history of computers

First we will look at the historical development of computers, in order to give a general understanding of computers in general use today.

The first machine generally thought of as an electronic computer was the ENIAC (Electronic Numerical Integrator and Calculator) built by Mauchly and Eckert at the University of Pennsylvania using 18,000 valves for the purpose of ballistics calculations, now more than thirty years ago. This machine, ENIAC, began the first generation of computers, which all used valves, and lasted until 1958. The next computers replaced valves by transistors, and formed the second generation (1959 to 1964).

Twenty years ago, in 1964, the IBM 360 computer appeared. This unified the up-to-then separate concepts of scientific computer or business computer, and on the hardware front introduced the application, though admittedly only as a very first step, of integrated circuits (ICs). The unification and fixing of computer architecture by the IBM 360 was widely acclaimed, and the third generation began. The design of the IBM 360 is to be seen behind many typical computers in use today.

The next, fourth, generation of computers is considered as those which implement thoroughly LSI (large-scale integration) technology, and, looking at current computer technology, we might say that we are just entering the era of the fourth generation.

The fifth generation computers are expected to exploit VLSI (very large-scale integration) technology to the full, but will probably not appear until the 1990s (see Fig. 6). If VLSI technology can be used, the central processing unit of a present-day large-scale computer will be able to be contained on a single chip, so that computers will become possible whose power is now simply not imaginable.

Up to about the third generation, hardware was the principal problem in implementing a computer, but since that time other problems such as software have come to the fore. As mentioned in the previous chapter, software is very difficult to create, and requires a large amount of effort for even a small correction. There is also the difficult problem of ensuring that an investment in software is not lost by having to throw away existing programs. If the computer architec-

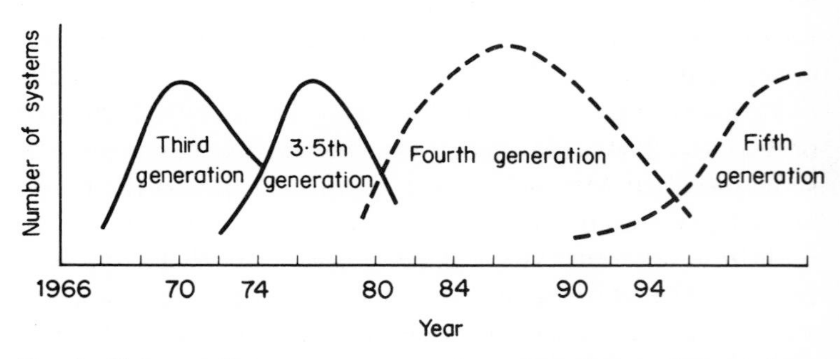

Fig. 6 Volume of computer system sales and evolution of generations (schematic)

ture changes, software must then also be changed accordingly. Similarly, changes in programming languages can have an enormous effect by requiring program rewriting.

In other words, as software accumulates, it becomes undesirable to change either computer architecture or programming language. From this point of view, therefore, the introduction of new technology begins to look more like a problem than a simple improvement. To a certain extent, it has been possible to preserve computer architectures more or less intact, and simply improve performance; thus the IBM 370 followed closely the architecture of the IBM 360, and now new machines are continuing this range, with the architecture substantially unchanged, at least from the user's point of view.

Von Neumann architecture

The mainstream of computers up to the present has been the so-called von Neumann architecture. At about the time that the first computer, the ENIAC, was built, a group under John von Neumann at the Institute for Advanced Study were considering the future model for computers. They considered new computer structures, and produced a collection of papers on the stored program type of sequential control, binary computer. Although the group later scattered worldwide, this idea was still the basis for computers built everywhere, and became known as von Neumann architecture.

How, then, did this von Neumann architecture differ from preceding computers such as the ENIAC?

The first difference is in the use of binary numbers, and the second in the stored program control concept. The concept of sequential control, which is described later, is a prerequisite for stored program control. The breakthrough concept of stored program control is in regarding both the program

Fig. 7 ENIAC: The first electronic computer

which describes the steps of the computation and the data used by the program as information, and in storing both types of information in the same place in the computer. This allows a program to be regarded as data and updated, and greatly increases the range of functions of the computer. Previously the program had been set up by some sort of hardware wiring, and the data expressed on a quite different medium. Stored program control made program input and updating very easy.

In the actual sequence of a computation, first the program which represents the sequence of calculations and the data required by the program are stored in the computer. Next, instructions are fetched sequentially from the store to the central processing unit, decoded, and acted upon to fetch data from the memory and perform calculations on it; the computation then proceeds cyclically. In other words the computer operates sequentially according to the specification of the program. This type of computer control method is called sequential control, and the basic idea of a von Neumann computer consists in the combination of these two concepts: sequential control and the stored program.

The von Neumann group were not the originators of the sequential control concept, however. The earlier Mark I computer at Harvard University and the ENIAC were also sequential control computers, but the program and data were handled as completely independent objects. The most important concept of the von Neumann group was stored program control with the data and programs treated the same.

The sequential control method is not the only one possible, however. For example, an analogue computer does not use a sequential control method at all. Typically, an analogue computer is used for solving differential equations, and an equation to be solved is represented by an arrangement of integrators and adders connected together. Compared with sequential control, this method is called simultaneous control.

Why sequential control?

Why then is the sequential control method adopted in practice? One reason is that the computer was conceived as a tool for numerical calculation. Before the computer era, numerical calculations were done on a tabletop calculator operated by winding a handle. To do a calculation, first the operator would be given a list of calculating instructions, and would then follow these while operating the machine, and writing down intermediate results in a sequential calculation chart. By repeating this process, the final results required would be derived.

If this job is simply to be given to a computer, this means building a calculating section (calculator) and providing something corresponding to paper on which to write the equations specifying the calculation and any intermediate results of the computation. The calculating section is the arithmetic unit of the computer, and the paper corresponds to the memory. Then the process of looking at the list of calculating instructions and carrying out the calculation by writing data sequentially on the paper is precisely the method of sequential control. In view of this resemblance, the computer structure seems quite natural.

Another reason for adopting the sequential control method was the necessity of building a system with as few components as possible. To build the first computers required an enormous number of valves—in the case of ENIAC, some 18,000. At the time, however, the average life of a valve was about 2,000 hours, so that in a machine containing 18,000, the average time between failures would be a little less than ten minutes. A valve that failed would have to be replaced immediately, and there was no way that such a machine could be envisaged working.

Something had to be done to improve reliability, or the number of valves had to be reduced. In fact ways were found in the circuitry to increase the life of the valves. Basically, in the design of the first computers the crucial questions were how to reduce the number of valves and bring about simplifications, so as to improve reliability. The adoption of sequential control was therefore essential to the realization of computers, by utilizing a single adder circuit to proceed iteratively through a computation.

The development of computers

When computers first appeared, few applications had been thought of other than in the preparation of mathematical tables and so forth. Therefore a very big effort went into producing a program to calculate the tables, and if the machine could simply print out the results without error it was deemed a success. At this stage the difficulty of programming was mainly in working out how to fit the program into a very limited memory.

As computer development has progressed, however, the range of applications has expanded enormously, and computers are now used not only for scientific and technical purposes, but also in business. With this development has come the requirement for a large body of programs. In the early von Neumann computers, every detail of a program, down to combinations of addition and subtraction had to be specified by someone, making the job of writing a program extremely time-consuming.

Therefore, in order to make programming easier, and to a certain extent automatic, programming languages such as FORTRAN and COBOL were developed. Such languages made large programs possible, but on the other hand the compilers which translate them into the machine code which the computer can understand are themselves big programs, and made larger memories essential.

At the same time, the early computers were expensive pieces of machinery, and for economic reasons it was important to utilize them fully. Methods were sought whereby the arithmetic unit could be kept busy all the time, so that the computer speed was not restricted by the speed of the peripheral input and output devices, and this is called batch processing. It is a method whereby a large computer has a number of satellite computers arranged around it for input and output purposes, so that the flow of work through the expensive central computer can be improved.

The need for a program, or operating system, to control and manage the overall running of the machine also grew up at about this time. This in turn increased the memory size required. Whereas, for example, on early computers 4K words (4096) was thought enough, compilers required upwards of 32K, and operating systems increased that amount to 256K or more.

Needless to say, as application programs get larger and more sophisticated, they too require increasing amounts of memory.

Problems in computer technology

As computers have developed, they have been subject to an unending flow of demands and criticisms. Firstly, there has been, for example, the requirement for a computer with a large memory and high speed. The arithmetic processing speed of a computer is certainly high compared to other methods of calculation, but in order to solve problems such as partial differential equations the speed is still not sufficient. Again, to solve problems involving the manipulation of large matrices, bigger and bigger memories are needed. The most urgent calls for faster processing and bigger memories have always come from applications related to scientific computations.

From the point of view of cost too, however, an individual calculation costs proportionally less to run on a large and powerful computer than on a small one, so that the economics of scale have motivated the desire for bigger and faster computers; indeed it might be said that this has been the main factor in the development of computer technology up to the present. All the same, the following are some of the problems which have come to light in computer technology as it has developed up to the present.

The first is that computers are oriented towards handling numbers, and their processing power in problems involving other types of data is very low. As mentioned in the previous chapter, when applied to a problem such as reading a handwritten character or processing images, even an extremely large and powerful computer will be unable to do jobs which would be trivial for a human, or will only be able to do them by using enormous amounts of time. Machine translation is another such example.

The second is that since computers are controlled sequentially, following the fixed order written in a program, when applied to a problem with intrinsically a very high degree of parallelism, such as the solution of partial differential equations, they can take an incredibly long time. Various methods have already been suggested to overcome this problem. For example, by arranging a number of processors in a matrix pattern, each carrying out its individual processing, we can provide a so-called parallel computer. A total solution is still to be found, however.

The third problem is the limit to expansion of scale. As described above, systems are getting ever larger, and the amount of memory occupied by the operating system, and the proportion of the time spent by the computer in managing itself, are also growing rapidly. There seems to be a limit to the size of the system if it is to be used efficiently, and there also appear to be limits on the extent to which ever-larger systems can provide better cost/performance ratios. On the other hand, advances in LSI technology will improve the ratios of mass-produced small-scale systems, and will reduce the need for larger systems. The benefits of distributed systems, improvements in system reliability, and so forth have also led to reconsideration of the virtues of large systems.

The fourth of these problems is that the architecture of current computers is designed on principles laid down in an age when hardware components were very expensive indeed, rather than being designed for ease of use. Once components are as easy to use as they now are, it has become possible to ignore conventional architecture, and consider new architectures adapted to the fields of application and methods of use. Better software productivity is needed, and architectures that make programs easier to write.

Supercomputers

The end result of the need for larger and faster computers in scientific applications is the supercomputer. General-purpose computers are in wide use in business and scientific applications, but there is a basic difference between these two applications, in that the latter generally requires complicated computation to be applied to a large quantity of data, and great efficiency can be expected from a computer constructed specially for this application.

There is no precise definition of a supercomputer, but we can think of it as an ultra-fast machine for scientific calculations. The supercomputer for high speed scientific computation will be an essential machine in the 1990s alongside the more narrowly defined fifth generation computer for knowledge processing.

Large-scale numerical calculation was the original driving force behind the development of the computer, and is still calling for increases in computing power. Spurred by this, the development of high power machines for scientific computation has been carried on principally in America, but has not necessarily been economically motivated. There has therefore been no spread to a wide base of applications, which have been largely limited to military and special scientific areas. This has hindered the development of supercomputer software and application technology, and perhaps even held back the development of supercomputers themselves.

The decrease in hardware costs over recent years has made it more economical in a growing number of cases to replace experiments by computer simulation, and this trend can be expected to be seen in more and more fields in future. This is linked to the reconsideration of supercomputers for commercial exploitation. Reduced hardware cost alone will not get the supercomputer accepted in a wide range of fields, however. In addition, there are many problems yet to be solved in the research and development of algorithms, programming languages, architecture, and component performance improvement, and only when these are overcome is it likely that the supercomputer will be universally accepted. In the next section we trace the historical development of supercomputers, and review some of the technological problems.

The development of supercomputers

The demand for larger and faster computers has been strong since their inception, and from a relatively early point designers began to be aware of the limits to component speed imposed by the speed of light. It is not really surprising, though, that this point was more strongly felt in the age,

Fig. 8 LARC (Livermore Automatic Research Calculator)

before integrated circuits, in which circuits were constructed from discrete components.

For this reason, by 1960 in the development stage of the IBM Stretch and Univac LARC computers, research was being done into advance fetching of instructions, so that the instruction fetch and execution cycles could be executed in parallel; these concepts were widely used in later computers.

One computer which received high acclaim for scientific and technical computing was the Control Data CDC 6600, which was first shipped in 1965. Up to ten arithmetic units were provided in the processor, so that about five times the processing power of any existing computer was obtained. Following this line, the next improvement led to the CDC 7600, which provided about five times the power again, using discrete components, but with a clock rate of 27.5 nanoseconds (1/1,000,000,000th of a second). This probably represents the limit for this technology.

Within a computer there are many elements functioning in parallel, and they are kept in step by synchronizing electric signals known as clock pulses. Since it is the basic time unit of a computer operation, the shorter the clock pulse, the higher the performance that can be expected.

The parallel processing concept

The first project in which a number of processors were arranged in a two-dimensional matrix, in order to be able to solve problems such as partial differential equations faster, was probably the SOLOMON project. This research system was sponsored by the United States Navy, and had 1,000 processors linked in the SIMD method.

SIMD stands for Single Instruction stream Multiple Data stream, and means that a single instruction is carried out for a large number of sets of data simultaneously. For example, there might be 2,000 data values sent in pairs to 1,000 processors, which would all carry out the same computation on each pair of values.

When, for example, solving differential equations, if the space over which the calculation is to be done is subdivided by a fine mesh, and the processing for each infinitesimal is allocated to a different processor, the processors can all execute the same processing, and a huge increase in power can be obtained by using the SIMD control configuration. Since the processors are arranged in an array, this type of computer is called an array processor. Using a thousand processors, it was anticipated that with technology from the early 1960s, a processing power of one hundred MFLOPS could be implemented.

This is a good point at which to discuss the ways of describing speed. Frequently speeds are described in terms of the rate at which a processor executes instructions, usually counted in millions of machine instructions per second or MIPS. Similarly, for scientific computation using floating point arithmetic, the unit is MFLOPS, megaflops, or million floating point operations per second, and also GFLOPS, gigaflops, for 1,000 MFLOPS.

The SIMD approach was actually applied to the Illiac-IV, which was designed and built on the budget of the United States Department of Defense, and was intended for seismographic analysis. Later it was installed at the NASA Aims Laboratory in Mountainview, California, and was employed in aerodynamics research. However, a combination of unreliability and the difficulty of actually using this type of computer structure, meant that achieving any results at all took a very long time, and as a result, this aspect of research was largely abandoned.

Problems arose with the Illiac-IV because it had its own very different architecture, at the same time as trying to use the very latest hardware technology. This should be a reminder of how important it is to proceed slowly, testing the way ahead at every stage. The main reason for the slow rate at which the system could be used was of course the difficulty of programming. The NASA researchers were later successful in applying the Illiac-IV combined with a high speed, very large memory, to the solution of the Navier–Stokes equation in fluid dynamics. This success is probably linked to the recent new NASA project.

Pipeline technique

A technique of high speed computation quite different from the parallel processing architecture of the Illiac-IV is pipelining, which has been investigated as a control method at various levels. In contrast to the SIMD configuration described above, in which a number of processors each carrying out the same operation are arranged in parallel, in pipeline processing, each process is broken down into a number of steps each of which is assigned to a separate processor; these processors are then arranged sequentially in a line.

We can think of this as akin to a factory production line, where each processor carries out its work on data passed to it from the preceding processor in the line. When that job is done, it simultaneously sends the result data on to the next processor, and receives data from the preceding processor to repeat the process. In other words data values are supplied at fixed time intervals, and after a number of processing steps, the result values emerge from the output end at the same time interval. Thus, in pipeline processing, at any particular time each processor is carrying out a different operation, so that it may be regarded as a kind of parallel processing.

Even in a sequential control machine such as the IBM 360/90, the pipelining technique is applied extensively to the internal operation of the arithmetic unit. One experimental ultra-high power computer using pipelining techniques is the Lawrence Livermore National Laboratories' STAR (String Array Processor) built by the Control Data Corporation with assistance from the American Atomic Energy Commission (AEC). Texas Instruments have also developed on their own behalf a system consisting of four pipelines called ASC (Advanced Scientific Computer).

All of these systems can produce very good performance when the length of a set of data to be processed is longer than the length of the pipeline, but in other cases often only a factor of about two can be gained over a sequential control machine.

Future supercomputer developments

Dr Seymour Cray, the designer of the CDC 6600 and 7600, abandoned design of the next model, the 8600, to form his own company Cray Research Inc. (CRI), and begin to develop a new computer, the CRAY 1. This built on the experience of the Control Data Corporation STAR and the Texas Instruments ASC, and can process at high speed even when the data volume is small; it was the motive force behind moving into the second generation of super-computer.

Currently operating supercomputers include, in addition to the CRAY 1, the Control Data Cyber 203 and 205, enhanced versions of the STAR 100.

There are a number of special-purpose systems for image processing which have had considerable acclaim; they are oriented towards processing in bit units, and include the Goodyear Aerospace MPP (Massively Parallel Processor) and the ballistic missile defence system PEPE. So far there has been no success in generalizing these systems to produce scientific supercomputers, but they will no doubt be an important subject for research as architectures for the application of VLSI technology.

One indication of future trends is that NASA have awarded a contract, which is in the early stages of design, to Burroughs Corporation and Control Data for a super-

Fig. 9 FACOM VP-200 supercomputer system

computer with a power of one GFLOPS, in order that wind tunnel experiments can be replaced by computer modelling. Independently, the Lawrence Livermore National Research Institute has its S1 project.

In Japan, Fujitsu and Hitachi have both marketed supercomputers, which are currently some of the fastest in the world (Figs. 9 and 10). Nippon Electric Corporation have also announced the development of a high speed machine.

In Japan there is also a large-scale eight- or nine-year project started in 1982 for a supercomputer which at ten GFLOPS will dwarf even the NASA project. This will also include basic hardware component development as a first stage. By advancing fundamental technology in algorithms, software, architecture, and hardware in step together, it is hoped to produce a supercomputer with good overall balance.

Fig. 10 HITAC S-810 array processor supercomputer system

Spurred on by this Japanese project, the American Department of Energy has announced intentions to encourage the development of a supercomputer, including the basic technology required, with a target power of 20 GFLOPS. With America putting its weight behind it, we can expect to see remarkable developments within the next ten years.

The fifth generation concept

Beyond all the computer developments and problems which we have discussed so far hope lies in the appearance of a new non-von Neumann computer, or fifth generation computer.

The concept of a non-von Neumann machine has been around since the early 1960s; for instance, for tasks in the artificial intelligence field, such as pattern recognition and machine translation, von Neumann computers have been considered incapable of providing sufficient power. The non-von Neumann computer has been proposed as a means of solving this bottleneck. The fifth generation computer is also targeted at solving the several problem points outlined up to now, and to making a reality of a new computer with a wider field of application and new methods of use.

In order to give a clearer impression of the fifth generation computer, we will now consider it from the point of view of the relation between the human user and the computer.

When two systems are exchanging information, the boundary between them must have a mutually agreed protocol for operations, known generally by the term interface. Conventionally instructions given to a computer have had to be in machine language, so that in other words there has been a machine language interface between human and computer hardware.

Machine language consists largely of very simple instructions such as add or jump instructions, so that there is a large gap in meaning between the sort of requirement that a human would feel and the machine language necessary to express it. It is the job of software to fill this gap, but this is very difficult to achieve at present because the gap is simply too wide. Taking FORTRAN as representative of the so-called high level languages, although it is at a higher level than machine language, and has been said to provide a second interface, there is still an enormous difference from requests expressed in human terms.

For example, procedural languages like FORTRAN are languages for specifying sequential processing, but not everything can be expressed easily in that way. In order for a computation to proceed, all details of the processing must be specified, down to the last exception, yet in no way can this be said to be easy to write. Thus there is still a considerable leap from the original human request to the completed program.

Ease of use

It is possible to imagine a programming language at a slightly higher level than the procedural languages such as FORTRAN, in which the user's request can be written: such a language is called 'non-procedural'. In a non-procedural language, the user has only to write the request which the computer is to obey, and leave to the computer the details of exactly what is to be done, and in what order. The computer must understand this language, and also have the ability to formulate a procedure for satisfying the request. From the user's point of view, it is not necessary to write the entire procedure, but sufficient to write simply what is required—a much more convenient matter.

In a conventional computer system information is stored in a database. A database can be thought of as a unified collection of various data items given a high level structure, so that processing can be carried out efficiently with any index, whereas the meaning of the data items is presumed to be known by the user and reflected in the program in such a way as to achieve the desired result. When a non-procedural language is available, then, when items of knowledge are required from a database, they will be obtained simply by writing that such and such data is required. It will not be necessary to write the procedure by which the data is to be sought. In order to implement this, then of course not only the data items but also their meanings must be stored.

In fact, though, in the human user's mind, the request

may initially be something more vague. Up to now, it has been the user's task, given a vague request, first to express it more lucidly in a non-procedural programming language, and then to convert it to a procedure-oriented programming language. If that task can be transferred only a little to system software and hardware, then the computer will be easier to use. For that to happen, system software must move further towards the human user, and new hardware functions will also be required.

Starting from this viewpoint, it is the basic concept of the fifth generation computer project to conceive of an image for future computers and to work on the research which they will require.

Of course, concepts such as these have probably existed for some time, but the reason for particular application to these problems in the fifth generation project at present is that several indications have appeared which suggest that these concepts may become possible. They will be discussed in detail in the next chapter, but the biggest single factor is the advancing emergence of new hardware component technology, centred on VLSI. There have been great advances in LSI technology, in price, in reliability, and in the density of integration, so that components of a scale and complexity which made them impossible previously, can be mass-produced at low cost.

Another important reason is that as research in artificial intelligence and software engineering have advanced, the reasons have become more apparent why computers have not so far handled problems such as artificial intelligence and parallel processing well, and methods of processing have also become clear. The possibility has appeared that, if research is continued, a computer can be built with a more natural interface to the human user.

Chapter 4

The basic technology reaches maturity

WE CAN EXPECT THE FIFTH GENERATION COMputer, which will make knowledge processing possible, to present a very different appearance from conventional computers, but various basic technologies, in which research has been making steady progress, will play an important part in its construction. For hardware implementation of high speed complex knowledge processing, VLSI technology and component technology are crucial, and also, at the system level, parallel processing technology is essential. Then for the implementation of computer networks which are usable over a large area, integration with communications technology is a central issue.

In this chapter we will discuss in outline basic technologies including VLSI, high speed component technology, communications technology, and parallel processing technology.

Development of VLSI

When a number of circuit elements are combined together on a silicon semiconductor chip this is termed an integrated circuit (IC), which can be classified according to the scale of integration as SSI (small-scale integration), MSI (medium-scale integration), or LSI (large-scale integration). At present, VLSI (very large-scale integration) circuits are being developed to increase the scale of integration even further.

With technological advances in the intricate processes for making components, and increases in the area of individual chips, the integration scale has increased up to the present at an annual rate of about two. VLSI technology has just begun to be used for memory devices.

A computer uses a memory built from RAM (random

access memory) chips—semiconductor devices which allow information to be read or written whenever required. RAMs can be further divided, according to the way in which the data is stored, into static RAMs and dynamic RAMs. The former retains the stored data as long as the power remains on, whereas with the latter type, the stored information disappears after a few milliseconds have elapsed. (A millisecond is 1/1000th of a second.) The data must therefore be repeatedly refreshed before it disappears; on the other hand, the power consumption of a dynamic RAM is less than that of a static RAM, which has the advantage of making a higher degree of integration possible.

Dynamic RAM chips currently in use have a capacity of 64K bits, and shipments of 256K bit RAM chips are starting. Since the structure of memory chips is more regular than that of processor chips, high integration levels are further advanced and, at the research level, laboratories have been successful in producing 1M bit RAM chips.

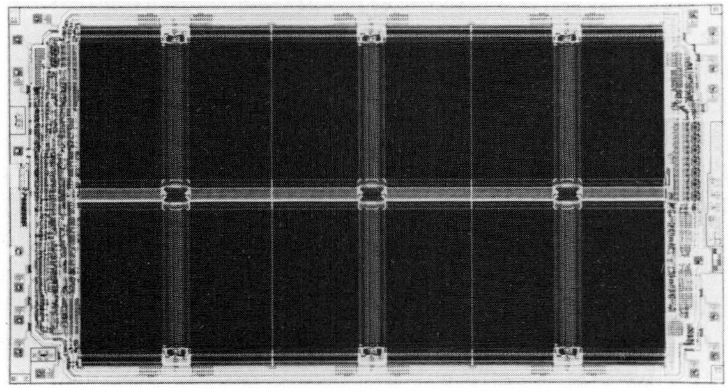

Fig. 11 256K bit dynamic RAM

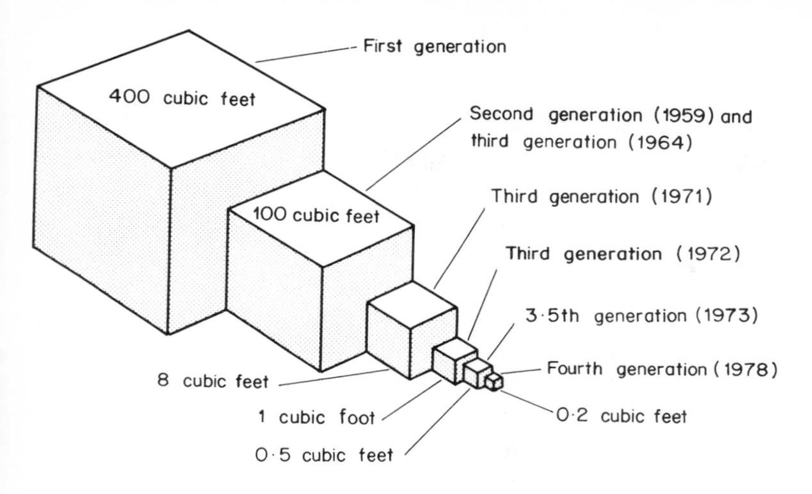

(1M = 1,048,576) Thus advances in the integration level of semiconductor memories are making steady progress.

The problem with applying VLSI techniques to processor chips is that, although the technology has been implemented to a certain extent in microcomputers, the stage has not yet been reached where VLSI technology can be applied sufficiently to general-purpose computers. If, however, semiconductor memory technology could be used in logic circuits, we could say that the technology has already been realized that would put upwards of 100,000 gates on a single chip, where a gate is the switching element for a single logic operation. In terms of the third generation of computers, this value of 100,000 gates means that the CPU of a large computer could be contained within a single chip.

This is a good opportunity to outline the basic structure and operation of a computer. Its main parts are a CPU (central processing unit), main memory, input–output devices, and secondary memory. Programs read in through the input–output devices are stored in main memory, and then executed sequentially by the CPU.

The secondary memory stores users' files, databases, and so forth, and this data is moved to the main memory as and when required, and then processed. The CPU, which is at the heart of the computer, includes control unit, arithmetic unit, and high speed memory (registers); the execution of instructions as described below is mainly controlled here.

First, the address stored in a control register called the program counter is sent to the main memory unit, and the instruction stored at that address is transferred to the instruction register. Next, this instruction is decoded, and data (operands) required for executing the instruction is obtained from the main memory unit. Finally this data is operated on in the arithmetic unit, and the results obtained are stored in registers or in the main memory unit.

The computer runs a program by repeating this operation for each instruction. A sequence of operations is directed by the control unit. The CPU can be regarded as the brain of the computer, and as such it is structurally very complicated. If this CPU were implemented in today's VLSI technology, it would fit into a chip of side less than one centimetre.

Design technology to match VLSI

An error-free system of this complexity on a single chip would be practically impossible with conventional human-designed technology. To apply VLSI technology to a logic unit, it is necessary to use the supporting computer systems which have been developed, known as VLSI CAD (computer-aided design) systems.

It is also necessary to consider just what computer

architecture is suitable for application to VLSI. For example, conventionally the complexity of a circuit has been interpreted in terms of the number of gates required. Thus, when designing parts of circuits with the same functions, designing for the minimum number of gates has been regarded as the optimum strategy.

In VLSI technology, however, the important question is not merely the number of gates, but also the length of the circuits interconnecting the gates, because space on a chip occupied by interconnections is effectively the same space as that occupied by the gates. Therefore, in a basis for design evaluation, the interconnecting circuit length must be taken into account in the same way as the number of gates.

This means that the best architecture for VLSI will have a structure with regularly patterned circuits alongside each other carrying out mutually interdependent processing.

Fig. 13 shows one example of such a structure. This is a computation method applied to VLSI called a systolic array, and developed by Carnegie–Mellon University in the USA; as shown in the figure, each cell is connected only to six neighbouring cells, and inter-cell communication is limited to these six cells. Cells of the same construction are arranged regularly in an array, so that their layout on a chip is extremely straightforward.

The purpose of this array is to calculate the product of two matrices: they are streamed in from upper left and upper right, and synchronized with this, the results of the calculation are output from the top. Data input and output and processing are pipelined, and the processing proceeds regularly in synchrony with the flow of data, like the heartbeat pulse from which the name is derived. Various structures have been studied in addition to this particular one, for their application to efficient layout on a two-dimensional plane.

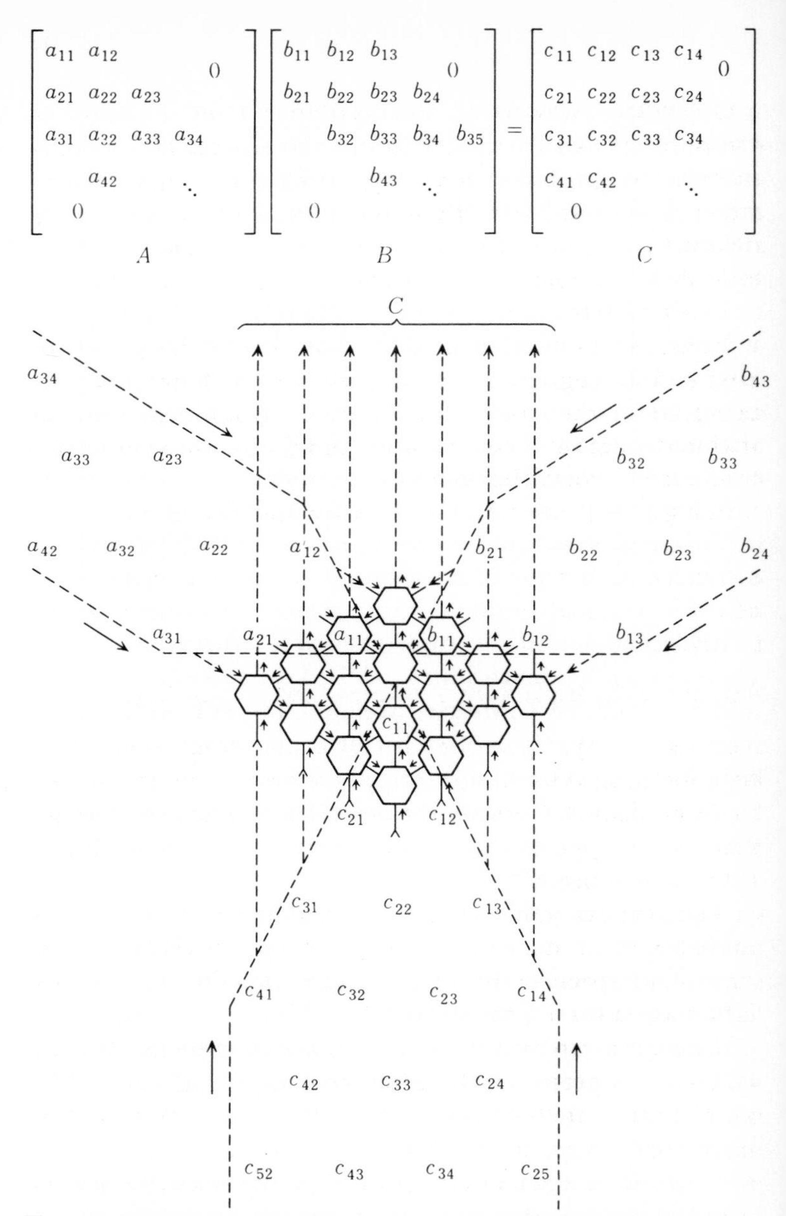

Fig. 13 Systolic array. Hexagonally connected array for calculating the matrix product $AB = C$ given above

One reason why we have been able to build a system as complicated as a computer lies in the adoption of a hierarchical design. In other words, rather than using transistors, resistors, and so forth, to build an indefinitely complex structure, they are thought of as combining on one level to form gates, and the gates are then combined to form a complex system.

In the hierarchical design concept, then, these gates are used to form registers, and implement basic functions such as addition, then these functions are combined into the arithmetic unit and control unit, and finally these units are combined to form the computer system.

Being able to use VLSI technology means that even more complicated systems can be designed, and therefore one extremely important problem lies in the development of new hierarchical design techniques to fit in with VLSI technology.

Advances in high speed component technology

Needless to say, up to the present faster components have been the central technology supporting advances in computer technology. Of course, if a better component technology than silicon appears, it will be adopted in future computers.

Even with present silicon technology, however, there are gates which can pass a signal with a delay of a few hundred picoseconds (1 picosecond = 1/1,000,000,000,000th of a second). Expressed in other terms, in 100 picoseconds light travels three centimetres.

If we speed up by a factor of ten, and assume an element which can operate in 10 picoseconds, in this time light travels three millimetres. Such a device is meaningless unless the computer itself is very small, and in terms of systems, this suggests that the limit to improvement in performance from higher component speeds is in sight.

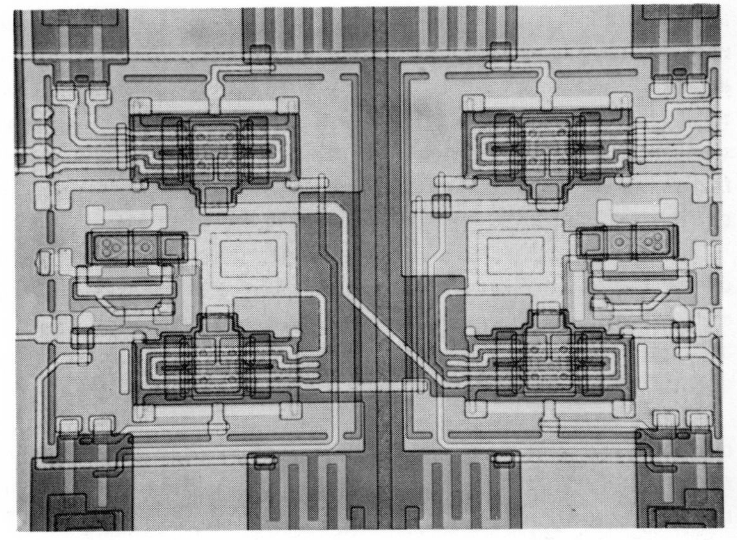

Fig. 14 Josephson device. Logic cell of logic gate array

At the same time, we may expect to see an effective improvement of between ten and one hundred times in the speed of processing elements, and such elements will include Josephson devices, HEMTs, and gallium-arsenide FETs.

Josephson devices use the Josephson effect in superconductors, and should provide extremely high speed operation, but require cooling with liquid helium (−269°C), so that there are many problems in developing practical devices.

HEMT is an abbreviation for high electron mobility transistor, a new very high speed transistor element. It uses a crystal heterojunction between the compounds aluminium-gallium-arsenide and gallium-arsenide. The ease of electron movement within the crystal is called the electron mobility; in a HEMT the electrons travel through the pure gallium-arsenide above the heterojunction, and therefore even at room temperature the mobility is about twice that of a gallium-arsenide FET, and at the temperature of liquid nitrogen (−196°C and below) it is about ten times. Since it can be cooled with liquid nitrogen such a device gives hope for practical implementation of high speed operation not significantly inferior to Josephson devices.

While a HEMT makes high speed operation possible, chemical compound semiconductors have an unstable sur-

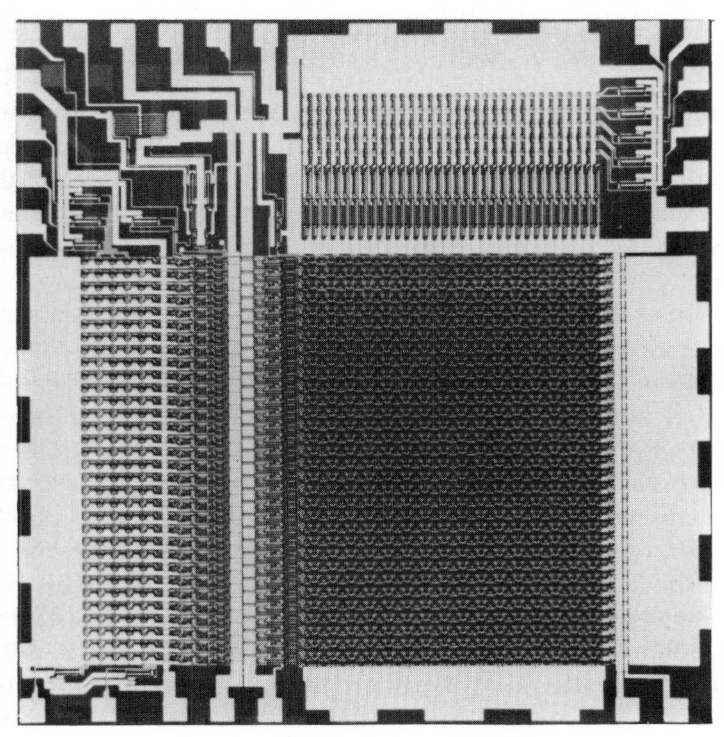

Fig. 15 HEMT. 1K bit static RAM

face, so that there are problems in the manufacture of dynamic RAMs with a high degree of integration.

In the last few years, there has been increasing interest in developing compound semiconductor gallium-arsenide integrated circuits to improve on the performance of conventional silicon components. A gallium-arsenide FET (field effect transistor) has the advantage over a HEMT or Josephson device of working at room temperature, and represents the technology closest to practical implementation. It falls short on speed, however, and promises only about an order of magnitude over present silicon technology.

Prospects for optical technology

In addition to these technologies, we must also look at optical technology. To overcome the limits on electronic signal transmission speed, some research has been carried out on the optical computer, which will process information using optical signals. However, using light in place of electricity, and producing logical elements with better characteristics than electronic devices, still requires a large amount of research and development, and such technology cannot be expected to be available soon. Of course, as optical communications technology shows, there are many advantages to be gained from using optical signals, and we can expect to see optical technology being adopted for various purposes in computers.

Input to a computer from the outside world is frequently in the form of an image, in other words optical data distributed over two dimensions. Thus light is a very suitable medium for transmitting a quantity of spatially distributed data in parallel, and optical technology may become important in application to parallel processing of data at the input and output level.

Integration with communications technology

The computer was developed principally as a data processing tool, but in an increasingly information-based society, it is important also to consider the information management role.

One objective of an information-based society is to enable data from all over the world to be captured accurately, and for the latest data from anywhere in the world to be available. This will require a fusion of computer technology with communications technology in a worldwide communications network to be constructed for data management.

Up to the present, communications technology has developed around the telephone, but that requirement is largely satisfied, and the growing need is for voice, image, and other data transmission. As a rapid conversion from analogue to digital technology is made, the interrelation of computer technology with communications technology will grow more and more. With advances in integrated circuit technology the costs of hardware have fallen dramatically; at the same time, digital technology has had an enormous effect on communications networks as circuits become smaller and more reliable.

The change-over to digital techniques has also brought suggestions for the setting up of a unified overall network to transmit and process different forms of data, and rapid moves are being made towards the international network architecture standardization which is necessary to ensure that various computers and terminals can be linked together in a standard way. If the communications functions can be standardized, then there is easy access to various resources on the network: supercomputers, databases, and so on.

If this can be achieved, then rather than processing on a

particular single computer, the various resources of the network will be available for distributed processing; technology must be established for this, and for the distributed databases which will be necessary for distribution of data among the computers on the network.

If data spread among geographically remote locations is to be handled as a unitary, consistent database, then there will be new problems not found in conventional databases, such as duplicated data and simultaneous update, but research on these topics is also proceeding. This still leaves many matters unresolved, such as the problem of security, but the fundamental technological necessity for a basic information-based society is the integration of communications technology with data processing technology.

Technology for parallel processing

While the final limit to component speed is coming into view, with the development of VLSI technology it has become possible to build complicated systems using many component elements. In order to implement a high power computer, therefore, we need technology for parallel processing so that a large number of similar elements can be organized to work in parallel.

If the data to be processed is in a regular form such as a vector or matrix, then as described in Chapter 3 under the heading 'Supercomputers', techniques such as pipeline control and array processor technology can be used, and very high power processing can be obtained. When, however, we have a mass of complicated, interrelated data, in which each part affects other parts, parallel processing cannot be implemented by application of pipeline or array techniques, so that it is necessary to reconsider the system structure from another viewpoint.

The MIMD type multiprocessor system, in which parallelism is implemented at a higher level, such as in procedures or tasks, has also received attention, and may be expected to be more widely used in future as low-cost VLSI processors are adopted. The big problem here is in determining how to implement most efficiently inter-processor communications, and again there is much research going on into methods of processor coupling. If the data transfer path between two processors is very large and powerful it is called a tightly coupled system, and otherwise, a loosely coupled system. The tighter the coupling the higher the power, but correspondingly the higher the cost of the coupling.

To build a large-scale multiprocessor system, requires that the coupling cost is limited, and suitable communications power is provided. When a group of processors is linked by a single bus, it is not possible to connect more than a small number of processors, and if a matrix switch is used the cost of the switch is very high, so that it is difficult to make a large system of more than a few hundred processors. Recently research has continued into hierarchically structured buses, cubic coupling and so on, and great insight is being gained into large-scale parallel processing systems.

Data flow architecture

From the system structure viewpoint, the data flow concept has received attention as a means of implementing parallel processing techniques in a natural way. J. B. Dennis and co-workers of MIT explain the basis of data flow architecture as data-driven and demand-driven.

The term ‘data-driven’ denotes the concept of control whereby an instruction may be executed just as soon as the necessary data is ready. Compared with the conventional computer in which instructions are executed sequentially, this new method means that an instruction is prompted into action by the arrival of the data which it requires.

In general, at any time, there will be many instructions ready for execution, and if there are a number of processors, several instructions can be processed at the same time, and we may expect faster computation than with conventional sequential computers. In addition, whereas until now for parallel processing it has been necessary to specify which parts of the program are to be executed in parallel, with the data flow method of control this is no longer necessary. This natural way of implementing parallel processing is seen as an important point, and its advantages are of course receiving attention in the fifth generation computer development project.

The control concept of 'demand-driven' means that the prompt to execute an instruction comes when the result of that instruction is required. Thus, when data necessary for one instruction is missing, a request for that data is sent to the instruction which produces it. Under this method, execution is limited to the minimum necessary, and wasteful computation can be avoided. As with data-driven control, many requests may be produced at once, and it is therefore suitable for parallel processing. When the result is returned to the point where it was requested, it is handled by 'data-driven' processing.

A computer built on these principles, and particularly using data-driven control is called a data flow machine. The data flow machine has a structure adapted to parallel processing, which is quite different from the von Neumann architecture, with its program counter and linearly addressed memory.

Programming languages for parallel processing

In order to make good use of parallel processing, research is needed into languages adapted to writing programs which include parallel processing structures. In a distributed processing system, in order for many processors with their separate interacting tasks to be coordinated to carry out to achieve a common goal, a high level language is needed in which the task of each process and the inter-process communication can be written clearly.

In present-day operating systems, several processes are executed in parallel on a single processor, in a shared-memory environment, but with the recent fall in microprocessor costs multiprocessor systems without shared memory have become possible; research has been done, by Hoare in the UK and Hansen in the USA, on parallel process description using message passing.

The description of a parallel processing program depends to a great extent on the architecture of the machine on which it is to be executed. In addition to languages for current general-purpose processor arrays, special-purpose languages have been developed, for example MPP developed by NASA for the SIMD image processing system. Again, research is also proceeding on languages for data flow machines.

In a parallel processing language, it is important that any parallelism that a problem possesses can be represented easily, but since programming languages are in the end only a tool, if the solution of the problem is intrinsically serial, nothing is to be gained by trying to impose a parallel structure.

Therefore, research on parallel algorithms for solving problems is very important. Human reasoning has a strong tendency towards the serial, and conventional algorithms are overwhelmingly serial, and largely unsuited to parallel processing. Algorithms with a high degree of parallelism are an important issue for the future.

Software technology

An enormous effort goes into the design, production, and maintenance of software, and this seems now to be the bottleneck in computer development. The cost of software development is outstripping hardware development costs, and this trend looks set to continue. The increase in demand for software is also rapid, and if it were to continue at the present rate, the point would soon be reached where even if the entire population of the world became programmers there would still be a shortage. This fact has led to much research, under the title of software engineering, into ways of increasing software productivity.

One software technique which cannot be overlooked is that of structured programming, pioneered by Professor Dijkstra at the Eindhoven Institute of Technology in the early 1970s. This advocates abstraction as a means of representing a problem in a more transparent manner, and underlined the importance of designing a program with a structure whose correctness can be easily determined. In the research which ensued, various means of improving program production were prominent, among them: modularity, abstraction, and data concealment.

Modularity means that a program is divided into modules as far as possible independent of each other, the interface with the outside is carefully specified for each module, and information is passed only through that interface; in this way complicated systems can be built more easily and with higher reliability.

By abstraction is meant the crystallizing of the essence of the problem in the basic structure of the program, ignoring details about representation. Various abstraction mechanisms have been studied, including abstract data type, in which the methods of access to data are restricted, and when the user wants to access data he must choose one of the methods defined on it. Thus the user cannot know the internal details of data, but also cannot destroy the data by use of inappropriate access methods. The program will therefore be easier to understand.

Data concealment is an extension of the idea of data abstraction, in which only necessary data is visible to the program, the remainder being hidden. The concept of 'object-oriented approach' takes this abstraction a stage further: both data and procedures are handled as objects, and all operations are expressed in terms of messages between objects.

In order to improve software productivity, these techniques are being introduced, but functional programming and logic programming languages, which are based on programming non-procedural models of computation, are also an important topic. Whereas in conventional programming languages a serial course of execution is described, in these new languages there is a static description of a set of definitions or facts. In addition to the ease of program writing, they are easy to understand. Typical examples are the functional languages LISP and APL, and the logic programming language PROLOG. The latter will be discussed further in the next chapter.

If these new methods can be properly utilized, then enormous increases in software productivity will be possible. Of course, though, as well as research into software itself, it is important that from a hardware viewpoint architectures are offered which allow efficient use of these new software techniques.

Pattern recognition

As the fields of computer application expand, the interface between human and machine must be as natural as possible in human terms, if an ever-wider set of users are to be able to treat the computer as a convenient tool.

This means the machine being able not only to understand speech, or to recognize characters, drawings, and images, but also to a certain extent to understand their meaning. This will require the machine to share some of the common knowledge of humans.

For the machine to approach human understanding, the important question is that of artificial intelligence. Since this is discussed in detail in the next chapter, here we will touch briefly on pattern recognition. We can cite character recognition as an example of a pattern recognition technique which is already in practical application. The Japanese Post Office mail automation system reads a handwritten numerical postal code, and other machines to read non-numerical characters are already being produced.

For printed characters, the technology for machine reading even of complicated Chinese ideographs is established. For handwriting, however, the situation is rather different, whatever the language. In general it will not be possible to read each individual letter with sufficient accuracy. Instead, what is required is an approximation to the human strategy of trying to understand the sentence, and deducing from the context any words which are difficult to read.

Even with a machine which can only read handwriting with say 90 per cent accuracy, if it knows that the writing represents an address, then given information about the structure of an address, and lists of names of cities, counties and so forth, it will be able to decode 99 per cent of cases. In other words, if the character recognition system can be combined with back-up knowledge, the recognition itself can be carried out very reliably. Much the same remarks apply to speech recognition or natural language understanding.

As to the extent to which speech recognition is possible at present, it has always been thought of as an extremely difficult problem, but recent advances have made possible an experimental system in which speech input is used for a word processor. At the same time, machines to discriminate ten words or so are already in use in bank accounting systems, and for a specified speaker, recognition is possible for almost normal conversation.

Of course, humans also have difficulty fully understanding speech, when for example listening to a song or a foreign language with which they are not too familiar. This must be because the method of human speech understanding is not simply a matter of recognizing individual syllables.

In other words, even without perfect hearing, humans understand what is being said from its content and context. For practical implementation of speech recognition systems, therefore, it will be necessary to incorporate artificial intelligence techniques.

Image recognition, and recognition of objects, present the same problems. To be able to pick out an image or object from a complicated background needs a satisfactory understanding of how humans recognize the object to be extracted.

Image recognition also provides interesting examples, though, where using image processing actually aids human image recognition. For example, diagnosis from chest X-ray images has always been thought a specialist medical task, but applying image processing can accentuate details so that even a laymen can easily detect a source of illness. This is a typical example where human intelligence is reinforced by the assistance of a machine.

When humans exchange information among themselves, they use various means including speech and images. For the exchange of information between human and computer to be natural from the human point of view, not only for speech but also for images, the computer must have the same kind of recognition ability through which imagery can be used as input to the computer.

Chapter 5

What is a fifth generation computer?

AS MENTIONED EARLIER, IT IS PREDICTED THAT IN the 1990s a wide array of computers will be utilized in a broad range of applications. There will be supercomputers for use in scientific and engineering calculations and large-scale simulation work, huge information systems interconnecting distributed databases by means of telecommunications networks, and microcomputers used as controllers for industrial robots and as structural elements of various types of systems.

The Fifth Generation Computer Systems Project is aimed at promoting the research and development of computer systems for use in the field of knowledge information processing. That is, whereas conventional computers can be thought of mainly as 'number crunchers', the FGCS Project is designed to pioneer new applications, especially in the field of artificial intelligence, and to create the computer systems needed in these fields.

Three systems comprising the fifth generation computer

Basic research in artificial intelligence (AI), a field that revolves around language understanding, has taught us a number of things. However, to conduct R&D on a practical scale in the field of AI, it is necessary to manipulate massive amounts of data. And if this data cannot be processed rapidly, then the work simply will not be practicable. With today's computers, even simple processing operations require a lot of time, a factor that has proved to be one of the bottlenecks to carrying out practical research work in the field of AI.

Moreover, even though research in AI requires the handling of massive amounts of data, basic research in this field still relies on the efforts of individual researchers. This makes preparing huge amounts of data difficult and, since present-day computers operate at such slow speeds, processing this data takes a long time. All this adds up to a very unpractical, inefficient process. For this reason, although a relatively good deal of basic AI research is being carried out, very few practical results are being obtained.

In other words, since conventional computers were designed primarily for processing numerical data, they are extremely weak when it comes to processing the non-numerical data which is so essential to AI. For example, today's computers are extremely limited when it comes to inference functions and/or association and learning functions.

The development of VLSI technology has brought with it an economical means of enhancing computer hardware functions. It should therefore now be possible to improve those hardware functions necessary to process non-numerical data and to perform inference processing. This is one of the goals the FGCS Project is striving to achieve. We call this hardware the 'inference processing and problem-solving system'.

Secondly, even conventional information systems must handle various types of data. This data is retained and utilized in the form of databases. The data stored in databases consist of numeric and character strings. The meaning of this data is understood by the database user, who creates a program based on that understanding in order to obtain the answer he desires.

However, a knowledge information processing system is designed to process knowledge itself. Such a system must store, in the form of information regarding the mutual interrelations between the various data items, the meanings that these data items have, and use these meanings in carrying out processing. Another of the goals of the Fifth Generation Computer System Project is to go beyond the processing of 'meanings' and implement a knowledge base that will make possible processing that takes into account the environment in which the problems exist. We call this the 'knowledge base system'.

Thirdly, it will be imperative that computers be made easy to operate in order to expand their applications and enable them to be used as tools and/or assistants by large numbers of people.

Ease of operation can be discussed from a number of standpoints. One example of making computers easier to operate is the standardized way keys are arranged on keyboard input devices. Programs are input into today's computers mostly by typing them in using standardized keyboards. However, if computers could be given instructions by voice, that is, by talking to them just like we talk to other people, then operating a computer would become a completely natural act, i.e. extremely easy.

In order to make it possible for humans to interface with machines in a manner similar to the way humans interface with one another when they exchange information, computers will have to be equipped with natural (human) language understanding capabilities. Another goal of the FGCS Project, therefore, is to realize an intelligent human–machine interface that enables humans to exchange information with machines in a fashion similar to the way they exchange information with one another, using voice (speech), pictures, and diagrams (graphics) all at the same time. We call this the 'intelligent human–machine interface system'.

What fifth generation computers will be able to do

Just what can users expect from fifth generation computers possessing functions such as those described above? The following is an attempt to enumerate these capabilities.

The first thing that users can expect from fifth generation computers is that they will be easy to operate. To achieve this, not only will the human–machine interface have to be 'near human' in nature, but the computer itself will also have to possess 'common sense', i.e. common knowledge, on a par with that possessed by humans.

When human beings converse with one another, we are able to understand each other, to each comprehend what the other wishes to say even if he/she does not express his/her ideas and thoughts completely. This is because we possess a similar degree or level of common-sense knowledge.

For example, when two Japanese are talking, and one says to the other, 'He is a college student, therefore he can read English', the listener understands the meaning of these words immediately. That is because both speaker and listener share the same fundamental knowledge, they possess the same common knowledge. That is, in order for a Japanese to enter college, he/she must have successfully completed junior and senior high school, and it is common knowledge that junior and senior high school students in Japan study English as one of their compulsory subjects. Without this shared common knowledge, even a simple statement like 'He is a college student, therefore he can read English', would be incomprehensible. This is where the difficulty lies in getting a computer to understand such a statement.

To achieve mutual understanding between humans and

machines through simple conversation, it will therefore be essential that a computer have a knowledge base containing common knowledge equivalent to that of human beings.

Other important elements of computer ease of operation, albeit from a different standpoint altogether, are low price and high reliability. And yet another vital condition for easy operation is standardized I/O functions that enable users to exchange information with computers in formats they are familiar with. One final condition essential to making computers easy to operate is user awareness of the functions possessed by the computer and the way it 'thinks' when it performs a job.

The second capability that users can expect from fifth generation computers is that they will gradually assume the monotonous, boring jobs now being done by humans, thus freeing the user to devote him/herself to more complicated and challenging tasks.

For example, systems that will assist humans in the decision-making process come to mind. Naturally, in the end it is the human who must make the decisions, but the computer will specify the types of data necessary to make those decisions, and/or perform the low-level decision-making task of actually selecting that data. This kind of system is well within the realm of possibility.

There is a movement to utilize computers to perform certain types of medical diagnoses, and a computer designed for this type of work can be considered another kind of decision-making system.

As already mentioned in a previous section, raising software productivity is a major task being assigned to today's computers. Therefore fifth generation computers will also have to contribute to this effort. There are many possible uses for such software development support systems. One typical example is enabling the use of non-procedural programming languages.

Up until now, programming languages have been procedural in nature, describing processing procedures in a sequential manner. However, when it becomes possible to use non-procedural programming languages to describe the contents that we wish to calculate and/or the data we wish to process, then it will be much easier to write programs. That is, when we have a problem we want to solve, non-procedural programming languages will enable us to provide the computer with only the specification of the problem, leaving the judgements and decisions regarding how that problem should be solved to the computer.

Another example of how software productivity could be improved is a system that makes it easy to re-use existing programs. If a computer system could be designed that was capable of implementing new software programs by combining them with existing programs and/or by simply altering a portion or portions of existing programs, this would greatly enhance software productivity.

It will gradually become possible to develop systems that will be able to locate existing programs that fulfil or come close to fulfilling the specification needed in a new program, and/or can automatically synthesize such a program.

AI research and the fifth generation computer

We intend to use the fifth generation computer to implement a knowledge information processing system (KIPS). Another way to describe a knowledge information processing system is as 'applied AI'.

AI can be thought of as the imitation of human intelligence by means of a computer, and is a field of study that came about because of an interest in finding out just what 'intelligence' is. Interest in 'intelligence' began to grow in the 1950s, and research was initiated to determine to what degree intelligence could be implemented on computers.

AI research began by using computers to prove mathematical theorems, to play games such as chess and draughts and to solve puzzles. It was at this time that interest in machine translation also began to grow.

However, because computer capabilities were still very low at that time, AI researchers were not always able to obtain satisfactory results. Because of this, most people thought that it would be impossible to use computers to implement anything approaching human intelligence.

Then, towards the end of the 1960s, basic research into human intelligence got a boost from the numerous places that began to manufacture prototypes of intelligent robots possessing 'eyes', 'hands', and 'brains', and from the start of widespread research into character, image and pattern (object) recognition.

One concept in AI research is the following. First, suppose we have a black box. Then, put a question to this black box and wait for it to answer the question. If you are unable, based on that reply alone, to determine whether a human being or a machine is in the black box, then that is artificial intelligence. In other words, the machine exhibits artificial intelligence, irrespective of the type of mechanism contained in the black box.

From the point of view of applied AI, this concept of a black box is fine. But another point of view may be taken when it is human intelligence itself that is the object of study, i.e. when we are looking at things from the point of view of the basic science of AI.

From the latter standpoint, interest is focused on trying to find out what kind of mechanism should be assumed in the black box in order for it to exhibit something approaching human behaviour. One approach to solving this problem is to do research on living organisms.

Another approach calls for using information processing models (computers) to simulate human knowledge and emotions. Investigating the degree to which this simulated knowledge and emotions represent those of real human beings then provides a basis for research into knowledge and emotions.

Thus there are two major fields of AI research, the field of cognitive science that satisfies interest in intelligence itself, and the field of knowledge engineering that puts artificial intelligence to practical use. Two objectives of the fifth generation computer will be to pioneer various new fields of applied AI through the development of knowledge engineering, and to extend basic knowledge concerning human intelligence via the research necessary to realize such new applications. Research into the fifth generation computer is progressing in line with these two objectives.

Following pattern recognition, the next higher level of AI research is something called 'Blocks World', an area of research first undertaken by Winograd and his co-workers. This type of research involves limiting the problem domain to a world created from wooden building blocks of various shapes such as bricks, pyramids, and simple boxes. Commands and/or questions are then put to robots, which must then respond correctly.

For example, researchers may instruct the robot to 'pick up the big, red rectangular brick!' and/or ask the robot 'What's in the box?' The robot will then pick up the specified block and/or move the appropriate block to 'see' inside the box in a manner indicating it 'understands' the researcher's command/question (see Fig. 16). That is, the information processing system (robot) responds just as if it understood the command or question put to it by the human researcher. Winograd and his co-workers have shown how to implement just such systems.

This has stimulated interest in tackling problems related to language understanding and the knowledge representation necessary to achieve that. This has been done by representing this knowledge procedurally, i.e. as a program.

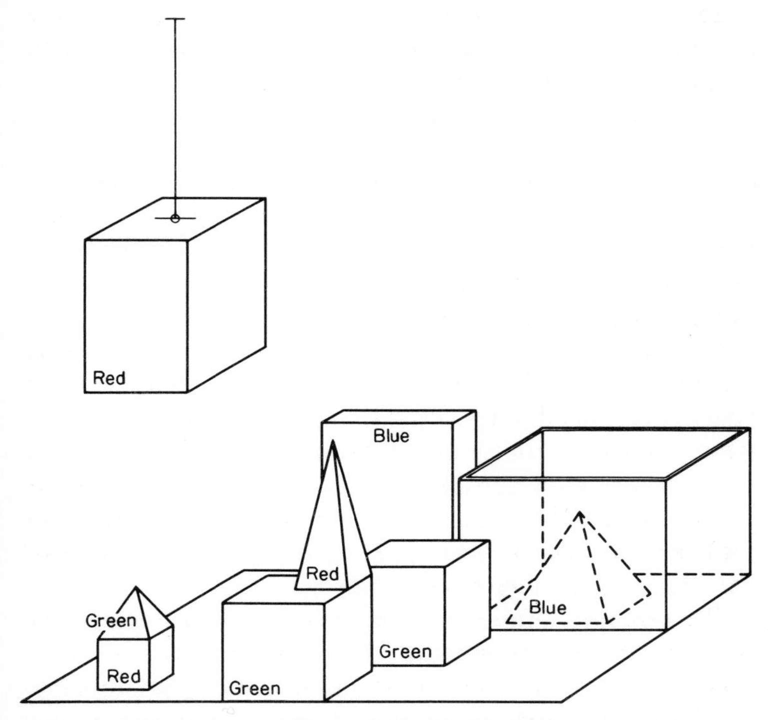

Fig. 16 Blocks World. When a robot is told to 'Pick up the big, red, rectangular brick', it behaves as if it understood the command

In order to get a machine to understand human language, it must be capable of: syntactical analysis, by analysis of the morphemes that comprise that language, and using the grammar of that language; analysis of the meaning of sentences; and context analysis, i.e. of the way in which the meaning may be affected by earlier or later sentences. The problem of determining just how to represent in computers the meanings expressed by the language in question is a difficult one.

As will be discussed later, various methods by which meaning can be represented have been proposed, involving the use of semantic networks and frames. Various hierarchical concept structures are also being considered.

We hope to use the fifth generation computer to create a system that is capable of using such methods to act in a manner closely resembling the intelligent actions of humans.

A general picture of the fifth generation computer system

Fig. 17 is a diagram showing the overall configuration of the Knowledge Information Processing System. This diagram is a general picture of a knowledge information processing system with the model systems representing the software system shown at the centre. The circle on the left side represents the world external to the computer, i.e. the user and external environment etc. The circle on the right represents the hardware system. The upper half of the model system in the centre, which represents the software system, is the intelligent programming system, and the lower half of the model system is the knowledge base system.

A user language employing voice or natural language, graphics, and images is envisioned as the interface between the user and the software system. As the diagram shows, this

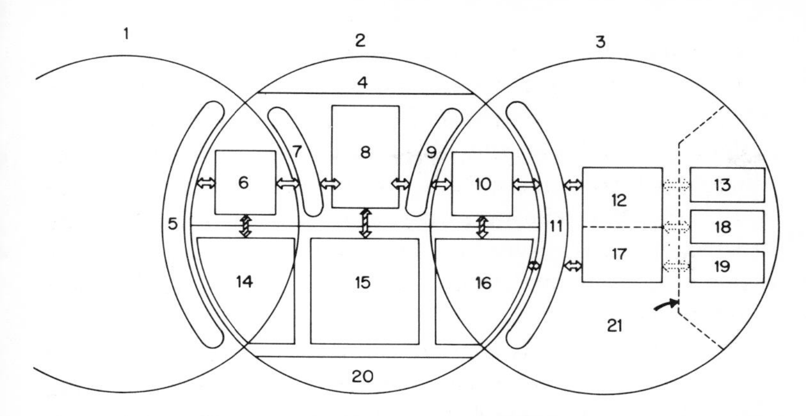

Fig. 17 Diagram illustrating the concept of the fifth generation computer system. (1) human system (applications system); (2) model system (software system); (3) machine system (hardware system); (4) intelligent programming system; (5) user language employing voice, natural language, graphics or images; (6) analysis, understanding and synthesis of voice and graphics etc.; (7) intermediate specification/response; (8) problem understanding and response synthesis; (9) processing specification/results; (10) program synthesis and optimization; (11) logical language/knowledge representational language; (12) problem solving and inference machine; (13) symbol processing machine; (14) knowledge about language and images; (15) knowledge about the problem domain; (16) knowledge about the machine system and knowledge representation; (17) knowledge base machine; (18) scientific computation machine; (19) database machine; (20) knowledge base system; (21) interface with fourth generation machines

user language is then analysed by the software system and converted into the intermediate specification. Knowledge of language and images will then be used to analyse voice and picture inputs.

The computer will understand a problem by looking at the intermediate specification that has been obtained by this conversion process. It will arrive at this understanding by using the knowledge (in its database) concerning the problem domain. Once the computer understands the problem, it will convert that problem into a specification for processing. This is indicated in the middle circle.

This processing specification will be used to synthesize and optimize programs using knowledge about the hardware utilized in the program synthesis system. Programs created in this way will be written in a logical language and processed by means of the problem solving and inference machine. The right-hand circle indicates this process.

The problem solving and inference machine will be supported by machines with symbol processing and scientific calculation function. A knowledge base machine will be required to fetch the required knowledge from the knowledge base. This knowledge will be described in the knowledge base machine by means of a knowledge representation language.

The results of this processing will then be used by the computer to synthesize its response to the user. The computer will again have to use its knowledge of the problem domain to synthesize its response. The diagram is also intended to show that the computer will synthesize voice and image output to the user by using its knowledge of language and images.

There are fourth generation computers with hardware functions such as symbol processing, scientific calculation, and database management. Therefore, the hardware that will be charged with performing these functions is shown in the area enclosed by the dotted line in the right-hand part of the diagram. The left-hand part from the dotted line in the diagram shows functions at present shared by software and humans.

The fifth generation computer will assign to hardware many of the functions currently being performed by software and humans. This will leave considerable room to manoeuvre within the software system. And this in turn will enable functions heretofore carried out by users to be relegated to the realm of software.

Therefore, the major software feature of the FGCS Project will be the utilization of a logic programming language as the system's kernel language. A logic programming language is one that embodies basic inference functions. The language called PROLOG (PROgramming in LOGic) is representative of such logic programming languages.

What kind of language is PROLOG?

Here we would like to present an example of a program written in PROLOG. The program consists of three parts: a collection of individual facts, general rules (knowledge base), and queries.

For example, the facts in this example program consist of two already known data:

(1) John and Mary are together;
(2) Mary is at Jane's house.

These facts are then augmented by rule (a), which states that 'People who are together are in the same place'. When we ask 'Where is John?', the system uses the two facts ((1) and (2)) and the rule (a) to arrive at the answer.

A special feature of PROLOG is that it allows the programmer to simply describe these facts and rules without having to concern himself with the procedure whereby the answer will be found. Our example program is shown in Fig. 18. The fact that Mary is at Jane's house is expressed: Location(Mary, Jane's house). And rule (a) expresses that if X and Y are together and Y is at Z, then X is at Z. The

```
Query    Location(John, ?)
Rule     Location(X,Z) ← Together(X,Y) ∧ Location(Y,Z)
Fact     Together(John, Mary)
Fact     Location(Mary, Jane's house)
```

Fig. 18 Sample PROLOG program (the ∧ sign is a logic symbol meaning 'and')

question mark (?) included in the query specifies the value being sought.

Fig. 19 shows how this program is executed.

When the query 'Location(John, ?)' is put to the system it searches its knowledge base to determine whether or not there are any facts or rules concerning 'Location'. When it finds the rule concerning 'Location', it tries the effect of substituting 'John' for 'X' (see (i) in Fig. 19). Now a rule expresses the fact that if the statement on the right of the arrow (←) holds, then the statement on the left of the arrow holds. The system therefore tries to solve the statement on the right of rule (i).

Since there is a fact concerning 'Together' which states: 'Together(John, Mary)', the computer tries the effect of substituting 'Mary' for 'Y' in this statement. This results in (ii). Now concerning 'Location', there is the fact that 'Location(Mary, Jane's house)'. From this, the system infers that the value of '?' is 'Jane's house'. In other words, the system derives the solution that 'John is at Jane's house' from facts (1) and (2) and rule (a).

The facts and rules used in this example program are extremely limited. However, ordinarily they are quite numerous, so that when it comes time to apply the rule(s), there are a number of possible candidates. When a solution cannot be derived from a certain rule, then the system tries each of the other rules in order (until a solution is found). As

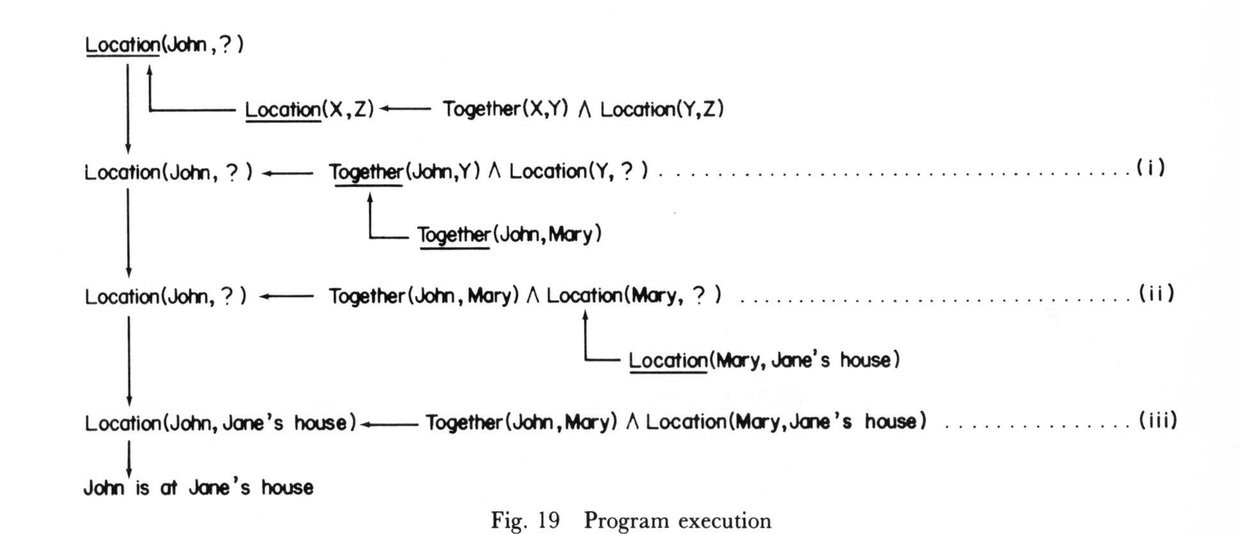

Fig. 19 Program execution

should be clear even from this simple example program, when a program written in PROLOG is implemented, the system has to perform a series of repetitious character string matching. This means that, as the number of rules and facts increase, the processing load becomes enormous.

Lightening the software burden

Using logical languages such as PROLOG very often enables programmers to describe human ideas quite naturally. Basically, this means they can simply write down the specification of the problem and then leave the processing entirely up to the hardware.

When the programmer writing a PROLOG program knows in advance just about what rules will apply to finding solutions to queries, then he can list the rules and facts in the order that they will be applied, thus creating a program that can be executed in a short time and run very efficiently. Present general-purpose computers that process data sequentially require more time to respond to queries the more rules and facts there are. Therefore, listing these rules and facts in their order of application when writing the program is very important.

If the time involved in executing the program is not of any great concern, then this type of control is not necessary. This frees the programmer from the tedious task of listing the specification in order, thus greatly reducing his workload. This in turn can be considered a means of enhancing program productivity.

In the era when computer performance was low and hardware costs were high, this type of program control was an important problem and a grave concern to programmers. However, today, with the steady maturation of VLSI and parallel processing technologies it is felt to be possible to reduce the burden of program control by means of introducing innovative computer architectures. That is, it has become possible to reduce the burden placed on software by transferring that burden to the hardware realm.

How is knowledge represented?

Before a knowledge base can be implemented, it is necessary to decide the format in which the knowledge will be represented. A number of representation methods have been proposed and are being researched, among them a 'frame' structure and 'network' structure. However, no single means of knowledge representation capable of being used in various applications has been definitely decided on yet.

One method of knowledge representation, called a semantic network, expresses knowledge in arcs that interconnect nodes. For example, the knowledge that 'A robin is a bird' is expressed by creating a node called 'bird' and another node called 'robin', and extending an arc from the robin node to the bird node.

This form of representation is simple and is a natural format as a model for human memory. But interpretations of the arcs and nodes can easily become ambiguous, and the generally large scale of these networks tends to make it difficult to decide what representation method should be used in the inference mechanisms based on them.

The frame model is a representation method based on Minsky's frame theory announced in 1975. One frame represents knowledge regarding a single object and event, the details and various related knowledge of which are described in a part of the frame called a 'slot'. By unifying frames hierarchically by means of their slots, a huge amount of knowledge can be organized into a frame system featuring a hierarchical structure.

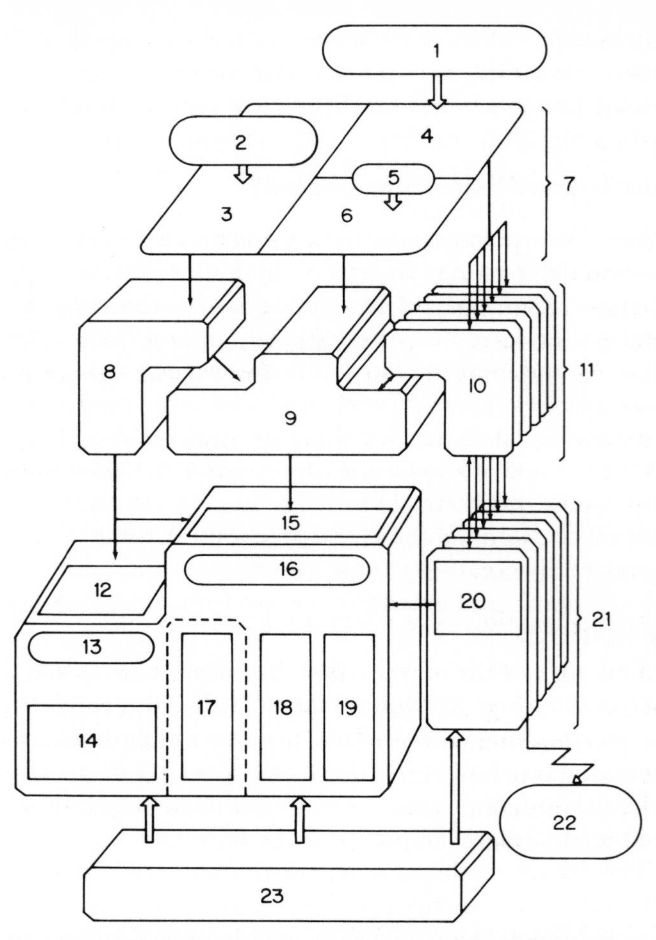

Fig. 20 Conceptual diagram of the fifth generation computer architecture. (1) access using natural language, speech and graphics etc.; (2) high-level enquiry language; (3) knowledge base management system; (4) intelligent interface system; (5) kernel language; (6) problem-solving and inference system; (7) external interface with the basic software system; (8) knowledge base management system; (9) problem-solving and inference system; (10) intelligent interface system; (11) basic soft-

A number of knowledge base systems have been constructed based on the frame model, but there are still a lot of points that are unclear concerning this representation method.

Deciding on a standard knowledge representation method is an important research task. The intention is to implement knowledge representation on what is known as a 'relational data model'. In a relational data model a clear-cut language interface called relational calculus, which is based on first-order predicate calculus, is established. Such relational calculus corresponds favourably with logical programming languages.

Relational algebra, which is the programming language that will handle the relational data model, has essentially the same descriptive power as first-order predicate calculus. This will make it easy to build a knowledge base machine on top of a relational database machine.

Fifth generation computer architecture

The concept of the fifth generation computer architecture is presented in Fig. 20. This diagram shows that the fifth generation computer will consist of three types of machine called the intelligent interface machine, the problem solving and inference machine and the knowledge base machine, and will be built using VLSI architecture.

---

ware system; (12) knowledge base machine; (13) relational algebra; (14) relational database mechanism; (15) problem-solving and inference machines; (16) predicate calculus type language; (17) abstract data type support mechanism; (18) data flow parallel implementation mechanism (data flow processing mechanism); (19) innovative von Neumann mechanism; (20) intelligent interface machine; (21) hardware system; (22) distributed function network system; (23) VLSI architecture

Because the system will be built using VLSI, the architecture used should be VLSI-oriented. In order to implement a VLSI-oriented architecture, it will be necessary first to use a structure in which the same VLSI can be used in large numbers repetitively, and second, to ensure that signal propagation in these VLSI chips is regular and short-distance, and that the interconnection circuits are short. The systolic array shown in Chapter 4 is one example of this.

Parallel processing should therefore be carried out at many levels to the maximum extent possible. In realizing the inference machine it is intended to adopt data flow systems as far as possible, since they can exploit in a natural way the parallelism that is inherent in various problems.

In order to represent a large program consisting of a combination of basic programs it will be very important to utilize an object-oriented architecture.

As already pointed out in Chapter 4, techniques for modularization and for abstraction will enable software productivity to be greatly improved. The idea of object-oriented architecture is also fundamental to these concepts. This means building an integrated system by means of objects that are equivalent to all the respective data and the operations that can respectively be performed on that data. We are therefore pushing forward with research on an object-oriented architecture, an architecture quite different from that used on conventional machines.

The knowledge base machine and the problem-solving and inference machine do not differ much from the standpoint of basic functions. The knowledge base machine is intended to derive necessary conclusions from a large number of facts, while the problem-solving and inference machine is designed to apply complicated rules to reach the necessary conclusions. Work on these two machines will therefore be integrated in future.

The intelligent interface machine will consist of various dedicated machines for respective media, such as a voice recognition machine, image-processing machine, and a graph-processing machine. As already pointed out, high-order understanding will require the use of commonsense knowledge in making inferences, which means that the problem-solving and inference machine and the knowledge base machine will have to work together cooperatively.

This section has presented an explanation of the concept of the fifth generation computer architecture and has covered the various functions of the fifth generation computer system, with emphasis on their high-level capabilities. However, it will also be necessary to provide low-cost machines with low-level capabilities as 'personal computers' for use by various individual users.

Therefore, we are also considering implementing distributed function systems that will enable users to select those capabilities available in each subsystem in accordance with their utilization objectives (processing needs) and configure the optimum system for each application. Super-machines will be constructed to carry out large-scale inference operations, and large-scale knowledge base machines will be built to handle large amounts of knowledge.

These various machines will then be interconnected via communication lines to realize a total fifth generation information processing system.

Chapter 6

The future of ICOT

IN 1982, THE JAPANESE GOVERNMENT'S MINISTRY OF Trade and Industry (MITI) began a 10-year project to make the fifth generation computer. The 'Institute for New Generation Computer Technology' (abbreviated to 'ICOT') was established on 1st April 1982, as the central organization responsible for the project, and research and development was begun in June of the same year.

This chapter will outline the research and development plan on which ICOT is based, and will also touch upon the influence that the fifth generation computer will have on future society.

Organization of ICOT

The main participants in ICOT are the eight computer manufacturers Fujitsu, Hitachi, NEC, Toshiba, Mitsubishi, Oki, Matsushita, and Sharp. Nippon Telegraph and Telephone (NTT) has also cooperated in this organization. A research laboratory with about forty young researchers working on the development of the fifth generation computer has been set up within the organization. These researchers are on secondment from the above-mentioned computer manufacturers, the Electrical Communications Laboratory of NTT, and the Electrotechnical Laboratory (ETL) of the Agency of Industrial Science and Technology of MITI.

The former Department Chief of the ETL Pattern Information Department, Dr Fuchi, holds the position of Chief of the research laboratory, having resigned from his position at the ETL and transferred to ICOT. Apart from the Chief of the laboratory, all of the other staff of ICOT are from one or other of the eight above-mentioned companies, the ETL, or the Electrical Communications Laboratory. Except for those from the ETL or the Electrical Communications Laboratory, all of the new staff are up-and-coming researchers who were less than 35 years old at the time they were seconded.

The research laboratory is organized into three sections, one concerned with hardware, one concerned with software, and one concerned with the rapid development of concrete computer systems as a tool for research purposes.

Research and development plan

The fifth generation computer research and development project is a ten-year program planned as three phases: an initial phase of three years, an intermediate phase of four years, and a final phase of three years.

Fig. 21 ICOT Research Laboratory: research being carried out using many terminals

Fig. 22 ICOT Research Laboratory: small group discussion in the conference room

The initial phase consists in accumulation, evaluation, and reorganization of the results of the research done up to now in the field of knowledge and information processing. With this accomplished, the fundamental research necessary for the development of a computer system, or research concerning the elements that will make up such a system, will be conducted.

In the intermediate phase, the results of the research in the initial phase will be modified, improved, and expanded. At the same time, efforts will be directed toward combining the individual elements and constructing practical inference subsystems and knowledge base subsystems.

The final phase will focus on the construction of a total system bringing together all of the subsystems, improving

Initial phase: Development of fundamental technology (approximately three years)

Intermediate phase: Subsystem development (experimental small-scale subsystems)

Final phase: Total system development (prototype system)

1. Modules for each functional mechanism for inference subsystems
2. Basic mechanism of parallel inference
3. Data-flow mechanism
4. Abstract data type mechanism
5. Simulator for operational tests
6. Application of VLSI techniques
7. Modules for each functional mechanism for knowledge base subsystems
8. Basic mechanism of knowledge base
9. Parallel relations and knowledge computation mechanism
10. Relational database mechanism
11. Simulator for operational tests
12. Application of VLSI techniques
13. Fundamental software system
14. Problem solving and inference software module
15. Knowledge base management software module
16. Intelligent interface software module
17. Intelligent programming software module
18. Pilot model for software development
19. Hardware of serial inference machine
20. Software of serial inference machine

1. (Inference subsystem)
2. Intelligent interface software
3. Problem solving and inference software
4. Inference mechanism
5. Intelligent programming software
6. Intelligent interface software
7. Knowledge base management software
8. Knowledge base structure
9. Intelligent programming software
10. (Knowledge base subsystem)
11. Experimental use basic application system
12. Parallel software development system
13. Intelligent interface hardware
14. Intelligent interface hardware

1. Inference and knowledge base mechanism (realized by VLSI)
2. Fundamental software
3. Basic application system software

Fig. 23 Research and development plan

techniques, reorganization, and dealing with the details of final objectives.

The steps involved in the research and development plan are illustrated in Fig. 23.

The fifth generation computer as discussed up to this point can be divided broadly into three parts, the problem-solving and inference system, the knowledge base system, and the intelligent interface system. Because research on intelligent interface systems has already been carried out by the individual computer manufacturers and NTT, the primary focus of ICOT is on the problem-solving and inference system and the knowledge base system.

The type of computer under discussion in the initial phase is a serial inference system. This will be necessary for the development of the software, which will be done later.

With respect to the parallel-type inference system (which seems the most promising prospect), in the initial phase, research is being conducted to determine how to make a system on the scale of ten units. In the intermediate phase, the production of a parallel system on the scale of 100 units will be attempted. When this is accomplished, in the final phase the production of a parallel system on the scale of 1,000 units will be attempted.

With respect to the knowledge base, the initial phase will focus on the relational database machine, and a serial-type data base machine will be made. In the intermediate phase, a parallel-type data base machine will be made, based on the previous machine. And in the final phase, the intention is to produce a knowledge base machine by combining the parallel database machine and the inference machine.

After ten years it is hoped that the prototype fifth generation computer which will be used to build the knowledge and information system can be made in such a way that it is generalizable both to ultra-high performance computers and to small-scale machines used as terminals. It is expected that individual enterprises will use this fifth generation computer as a standard for developing their own products.

A software system which has a variety of functions, such as retrieval of knowledge and acquisition of knowledge, and which provides the basis for performing problem-solving and inference with this computer will be required.

At the same time, in order to demonstrate the usefulness of this computer as a knowledge and information system, the production of a number of practical systems is being considered. One of these is a relatively high-level machine trans-

Fig. 24 The relational database machine, DELTA, developed by ICOT. It performs the storage, retrieval, and updating of large amounts of data very rapidly

Fig. 25 The problem-solving and inference machine, PSI, developed by ICOT. It analyses problems and considers how inferences should be drawn

lation system, and another is a consultation system involving a number of specialized fields.

Furthermore, realization of a software production support system is being considered. This support system would retrieve software of similar type that had already been produced, modify it, and make software with new purposes in order to compile automatically many kinds of software and to generate software systems efficiently.

It is also hoped that an intelligent VLSI CAD system can be realized by amassing the knowledge obtained previously

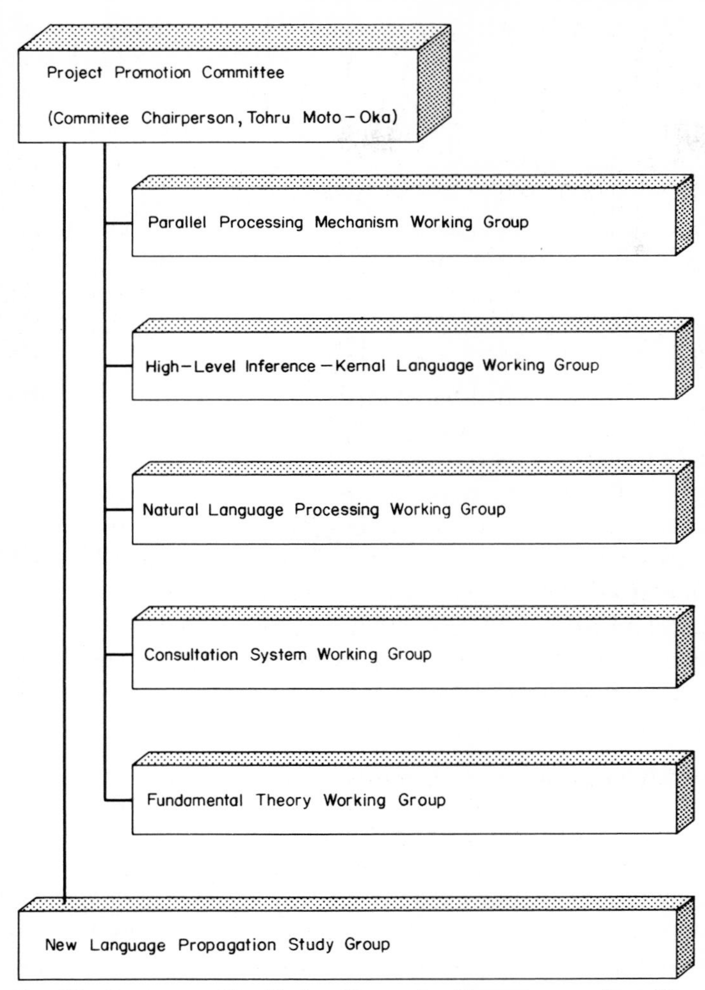

Fig. 26 Structure of the Project Promotion Committee and working groups

in VLSI manufacture and applying this knowledge to construct new VLSI.

In order to advance the aforementioned research, the project has been inaugurated with an estimated scale of funding of 10,000,000,000 Yen in the initial phase, 50,000,000,000 Yen in the intermediate phase, and 40,000,000,000 Yen in the final phase. The end of the initial phase is being approached and it has been estimated that about 8,500,000,000 Yen have been used. It is thought that the results of the research have been in keeping with the initial plan.

The fifth generation computer project is an extraordinarily difficult undertaking which includes a large number of new fields of research. Not only the government research laboratories and the public enterprises but the universities as well are contributing to moving the project forward. For that reason, a Project Promotion Committee was formed within ICOT. The structure of the committee is shown in Fig. 26. The specialists enrolled in the universities also participate in the individual working groups of this committee. They advise, participate in debates, and contribute to the progress of the research.

The fifth generation computer and society

The fifth generation computer is expected to be the leading type of computer in the 1990s. The fields of application of this computer will be extremely wide, and it will be used in systems to support many human mental activities.

However, a precondition for such a system is the creation of applications software appropriate to these fields of application of the computer. Just because a computer is completed, it does not mean that such systems will suddenly come into existence.

In order to produce systems which will be useful in each of these various fields of application, it will be necessary to create and develop projects as large or larger than the present fifth generation project in each field.

The fields in which it is thought that the fifth generation computer will be used are shown in Fig. 27. As this figure shows, the fifth generation computer will be used in the wide variety of mental activities that people carry out in offices and planning rooms in very many fields.

For example, let us take up the issue of the problems in education. The ideal instruction can be thought of as an educational method which corresponds to the capabilities and nature of each individual being educated. In the future, educational support systems using fifth generation computers will demonstrate considerable strength in providing such education directed toward the individual.

Of course, there are fields of education in which study can initially be performed by students in groups. In such fields and in fields such as instruction concerning aesthetic sensibility, the current form of education in schools can still be chosen, but we believe that computers will come to play an important role in education which takes the form of imparting individual items of knowledge. And they will also be used to give technical instruction within firms.

In translation too, systems which will support human operations will be created, and the fifth generation computers will come to replace a large portion of the translation currently being performed by people. However, it will not be possible for computers to do highly artistic translation such as that required for poetry and novels.

In the final analysis, basically it seems correct to think that machines to support most of the mental activities of human beings will be made in the future. The mental

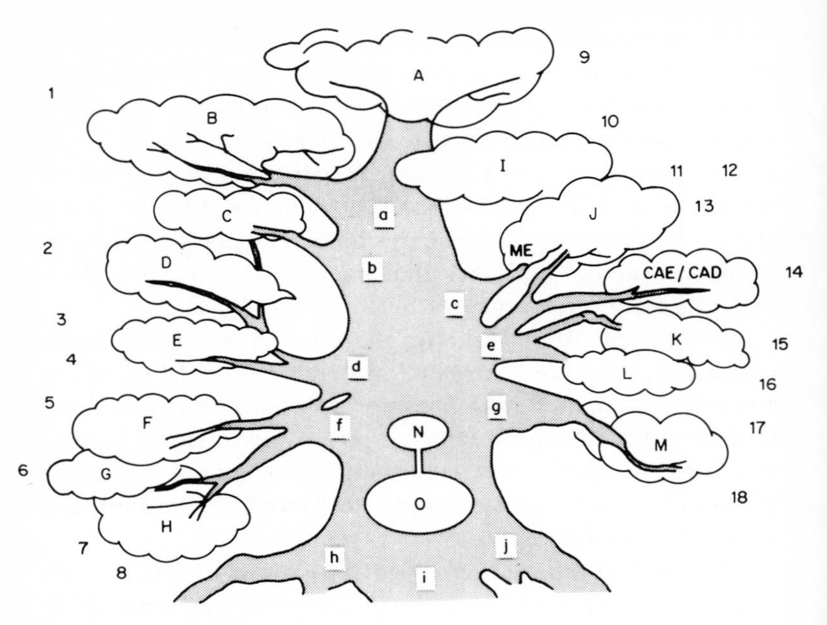

Fig. 27 Effects of the spread of the fifth generation computer.

(1) improving productivity in low-productivity fields; public services, offices, government, currency industry; (2) development of the database industries (acquisition of bargaining power); (3) elimination of international gaps; (4) accumulation of know-how; (5) increasing intellectual productivity in research activities; (6) creation of courseware; (7) home computers; (8) hobby computers; (9) making business decisions and refining government policies; (10) development of the knowledge industry; (11) decreasing costs of medical treatment and improving welfare; (12) development of health insurance industry; (13) solving the 'ageing population' problem; (14) improving the international competitiveness of the engineering industry; (15) improving productivity in manufacturing industries and conserving energy and resources; (16) diversification and customization of computers; (17) increasing productivity in the computer industry; (18) markedly improving the productivity of the software industry.

(A) decision-supporting systems; (B) advancement of office automation; (C) voice typewriter; (D) automatic compilation of databases; (E) automated translation; (F) super personal computers; (G) computers capabilities of people will be improved by these machines, and one may hope that a better society will be born.

If such computer systems progress into a variety of fields, it would be natural to expect that the nature of the work allotted to human beings in future societies will change as well. As a result of this, there are people who think that the society of the future will be controlled by computers. However, computer technology is ultimately only a tool with which humans have been provided. The question of how such tools are to be used is a new and challenging topic which people will have to deal with. The free time that is created by computers can be applied toward the progress of humanity and improvement of life.

The provision of these new tools will in fact extend human capabilities, enabling human beings to discover new possibilities for the future.

The question of whether or not powerful tools such as computers are a move in the right or the wrong direction for society is an important topic which humans are in charge of. The new possibilities with which we are provided, and the question of how to use them, require adequate discussion by a large number of people and must be examined carefully now and in the future. It is certainly true that this is an extremely important subject. But we believe it is the responsibility of mankind to strive to develop fully the

---

for use in individualized instruction; (H) Japanese language machine; (I) several kinds of expert systems; (J) medical treatment consultation system (ME); (K) CAM robot; (L) VLSI-CAD; (M) automatic program generation; (N) inference; (O) knowledge base.

(a) output of qualitative information and analogue information; (b) voice recognition; (c) picture recognition; (d) natural language processing; (e) picture processing; (f) intelligent interface; (g) intelligent programming; (h) Josephson GaAs; (i) VLSI (silicon); (j) optical fibres.

new possibilities that are given to us rather than to impede technical progress out of anxiety about its negative effects.

This is not limited to computers. The development of leading-edge technology in a number of fields will be promoted actively, and will have a great influence on future societies and cultures. The destruction of humanity which can accompany mistaken use of such leading-edge technologies is only too evident when one considers one of the applications of nuclear power.

Consequently, at the same time that leading-edge technology is being developed, it has become important to subject it to thorough scrutiny from various angles in order to make a society in which this technology is used in a wise and enlightened way. Simulation techniques using computers will no doubt be utilized to assist in such studies also.

Sources of the photographs and figures

Figs. 2, 3, 23, 24, 26, 27: Courtesy of ICOT.
Fig. 5: Courtesy of *Mainichi Shimbun*, Morning Edition, 21st June, 1984.
Fig. 8: Courtesy of Japan Univac.
Fig. 9: Courtesy of Fujitsu.
Fig. 10: Courtesy of Hitachi Manufacturing Works.
Fig. 11: Courtesy of NEC.
Fig. 13: *Introduction to VLSI Systems*, C. Mead and L. Conway (Translated by Takuo Sugano and Hiroyuki Sakaki.) (Baifukan, 1981, p. 305, Fig. 8.11 and Fig. 8.12.)
Figs. 14, 15: The Scientific Computer Research Association.
Fig. 16: *Artificial Intelligence*, P. H. Winston (Translated by M. Nagao and Y. Shirai.) (Baifukan, 1980; p. 165, Fig. 6.2.)
Fig. 17, 20: Proceedings of the International Conference on Fifth Generation Computers, October 19–22, 1981, JIPDEC.

Index